T0230525

Forum for Interdisciplinary Mathematics

Volume 3

The Forum for Interdisciplinary Mathematics (FIM) series publishes high-quality monographs and lecture notes in mathematics and interdisciplinary areas where mathematics has a fundamental role, such as statistics, operations research, computer science, financial mathematics, industrial mathematics, and bio-mathematics. It reflects the increasing demand of researchers working at the interface between mathematics and other scientific disciplines.

More information about this series at http://www.springer.com/series/13386

Ivaïlo M. Mladenov · Mariana Hadzhilazova

The Many Faces of Elastica

 Springer

Ivaïlo M. Mladenov
Institute of Biophysics and Biomedical
 Engineering
Bulgarian Academy of Sciences
Sofia
Bulgaria

Mariana Hadzhilazova
Institute of Biophysics and Biomedical
 Engineering
Bulgarian Academy of Sciences
Sofia
Bulgaria

ISSN 2364-6748   ISSN 2364-6756   (electronic)
Forum for Interdisciplinary Mathematics
ISBN 978-3-319-87032-8   ISBN 978-3-319-61244-7   (eBook)
DOI 10.1007/978-3-319-61244-7

Printed on acid-free paper

This Springer imprint is published by Springer Nature
The registered company is Springer International Publishing AG
The registered company address is: Gewerbestrasse 11, 6330 Cham, Switzerland

To the memory of my son Kliment who passed away so young, and to my wife Clementine for all her love with a hope that sometime, somewhere all of us will be together again and happy as before.

Ivaïlo M. Mladenov

To my husband Rosen and to my children Tsveta and Michail for their love and patience. And to Ivalo Mladenov for his support.

Mariana Hadzhilazova

Preface

This book has traveled strange. It began with an idea to provide a presentation of classical differential geometry that was as sufficiently complete and as rigorous as possible.

In principle, it was addressed to students or postgraduate students who are interested in the applications of differential geometry. This was fulfilled to a great extent, but two other powerful processes were in action simultaneously.

The first one, in a loose sense, was the economic crisis, which affected not only the financing of science in general, but also the level of training of students.

The second one was the parallel accumulation of many interesting results obtained in collaboration with our colleagues Vassil Vassilev, Peter Djondjorov, Vladimir Pulov, Borislav Angelov, Peter Chernev, Petko Marinov, Elena Popova, Jean-Francois Ganghoffer (Nancy, France), John Oprea (Cleveland, USA), and Jan Sławianowski (Warsaw, Poland). It is our pleasant duty here to express our gratitude for the cooperation we have enjoyed for so many years.

A part of these results was presented from a unified point of view in the Ph.D. thesis of the second author, defended in July 2012. At that time, it was decided that it should be expanded in the form of a book. Unfortunately, due to their administrative duties, our main co-authors (V. Vassilev and P. Djondjorov) could not take part in this new project.

The exposition itself is based on the natural mathematical framework which embraces the results mentioned above. Namely, the equation $\dot{\kappa}^2 = P_4(\kappa)$, where κ is the curvature of the plane curve and $P_4(\kappa)$ is a fourth-degree polynomial with real coefficients.

This polynomial appears in the most general case in considerations related to the shape of elastic cylindrical membranes under pressure, and the corresponding equation is accordingly called the equation of the generalized elasticas. It turns out that under the proper light, many other problems can be viewed as generalized elasticas. From this follows the book's title.

The reader has to decide for himself if we have succeeded in pursuing this aim.

Sofia, Bulgaria   Ivaïlo M. Mladenov
November 30, 2016   Mariana Hadzhilazova

Contents

Symbols

s	Archlength parameter
XOZ	Cartesian coordinate system in the plane
$\mathbf{e}_1, \mathbf{e}_2$	Basic vectors in the plane
$\mathbf{e}_1, \mathbf{e}_2, \mathbf{e}_3$	Basic vectors in the space
$\mathbf{x}.\mathbf{z}$	Scalar product of vectors
$\mathbf{x} \times \mathbf{z}$	Vector product of vectors
$[\mathbf{xyz}]$	Triple vector product
$\mathcal{C}$	Plane curve
$\tilde{\mathcal{C}}$	Evolute or deformation of the curve $\mathcal{C}$
$\mathbf{x}$	Point of the curve $\mathcal{C}$ or of the surface $\mathcal{S}$
$\tilde{\mathbf{x}}$	The point, which is infinitesimally close to $\mathbf{x}$
$\mathbf{T}$	Tangent vector to the curve
$\mathbf{N}$	Normal vector to the curve
$\mathbf{B}$	Binormal vector to the curve
(x, z)	Cartesian coordinates of the point in the plane
$(x(t), z(t))$	Parameterization of the plane curve
θ	The angle between the tangent vector and the OX axis
κ	Curvature of the curve
τ	Torsion of the curve
$\mathcal{R}$	Curvature radius
$\mathcal{D}$	Plane domain
Γ	Boundary of the plane domain
$\mathcal{S}$	Surface in the three-dimensional space
$\mathbf{x}_u$	Tangent vector to the u-line on the surface
$\mathbf{x}_v$	Tangent vector to the v-line on the surface
E, F, G	Coefficients of the first fundamental form
M, N, L	Coefficients of the second fundamental form
$\mathbf{n}$	Normal vector to the surface $\mathcal{S}$
$\mathbf{k}$	Curvature vector
$\mathbf{k}_t$	Tangential component of the curvature

$\mathbf{k}_n$	Normal component of the curvature
κ_n	Normal curvature
g	2×2 matrix of the first fundamental form
h	2×2 matrix of the second fundamental form
J	Functional
K	Gaussian curvature
H	Mean (average) curvature
Γ^i_{jk}	Christoffel symbols
κ_μ	Meridional curvature
κ_π	Parallel curvature
$\mathcal{A}$	Surface area or part of it
V	Volume of the body or part of it
am	Jacobian amplitude function
sn, cn, dn	Jacobian elliptic functions
F, E	Elliptic integrals of the first and second kinds
$K(k), E(k)$	Complete elliptic integrals of the first and second kinds
$\Pi, \Pi(n,k)$	Complete and incomplete elliptic integral of the third kind
$\mathcal{A}_C$	Cross-sectional area
$\mathcal{R}_u, \mathcal{R}_v$	Curvature radii of the u- and v-lines
Δp	The pressure difference on both sides of the membrane
$\Delta_\mathcal{S}$	Laplacian defined on the surface $\mathcal{S}$
$r_{\max}, r_{\min}$	max, min distance from the axis of symmetry
k	Modulus of the elliptic function
$\tilde{k}$	Complementary elliptical modulus
σ	Surface tension
Ih	Spontaneous curvature
$k_c, \bar{k}_c$	Elastic moduli
$\mathbf{v}_1, \mathbf{v}_2, \ldots, \mathbf{v}_6$	Generators of the Lie algebra of the Euclidian group
ℓ	Length of the curve
$\ell(P)$	Length of an extension
$\mathcal{A}(P)$	Area of one extension
$\mathcal{V}(P)$	Volume of one extension
ε	Deformation parameter

Abbreviations

ATP   Adenosine triphosphate
CMC   Constant mean curvature
C-M-P   Codazzi-Mainardi-Peterson
EPR   Electron paramagnetic resonance
G-C-M   Gauss-Codazzi-Mainardi
pH   pH indicator
RBC   Red blood cell

Chapter 1
Geometry and Variational Calculus

Abstract This chapter contains all necessary information in regard to the differential geometry of curves and surfaces needed for the description of membrane shapes. The main focus is on the geometry of the plane curves to which most of the considerations in the book boil down. As usual, coordinate systems, curvature and the Frenet-Serret equations are discussed in some detail. Then, space curves together with their tangent vectors, normal planes and curvature are introduced. Other important notions like principal normal and binormal, osculating plane and moving frame are introduced, exemplified and discussed as well. The Frenet-Serret equations for the space curves are derived and the principal theorem in the local theory of these curves is proved. The second part of this chapter is devoted to variational calculus, as this is the setting in which the problem of the optimal form of membranes is cast later on. Variational calculus occupies a special place in mathematics. Many of the laws of nature are formulated as variational principals. Hence, there is a great interest in the calculus of variations in many applied areas—from celestial mechanics to mathematical economics and management theory. Specifically in this chapter, the equations named after Euler and Lagrange are reviewed (and in some cases derived) in various settings, e.g., for functionals depending on one or more variables and comprising derivatives of first or higher order, etc. Most of these cases are illustrated by examples, which serve both to assist in understanding and to prepare the reader for the applications of the method to follow.

1.1 Plane Curves

Among many different coordinate systems used for presentation of plane curves, we choose only those two which are appropriate for our considerations. These are the parametric representations using the Cartesian coordinate system and polar coordinates. The coordinate axes of the Cartesian coordinate system are specified by a pair of perpendicular lines, the so-called abscissa (the OX axis) and the ordinate (OZ axis) (see Fig. 1.1).

The coordinates in the polar coordinate system are composed of a point (pole) and a vector from this point. Further on, we shall assume that the pole and the

© Springer International Publishing AG 2017
I.M. Mladenov and M. Hadzhilazova, *The Many Faces of Elastica*,
Forum for Interdisciplinary Mathematics 3, DOI 10.1007/978-3-319-61244-7_1

Fig. 1.1 Geometry of the plane curves

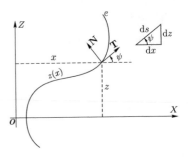

axis coincide with the origin of the positive part of the OX axis. In the Cartesian coordinate system XOZ, the coordinates of the curve $\mathcal{C}$ are given by two independent real functions of the real variable t in the form

$$\mathbf{x} = (x(t), z(t)), \qquad t \in I = (\mathring{t}, \bar{t}) \subset \mathbb{R}.$$

The functions $x(t)$ and $z(t)$ form a *regular* presentation of the curve (arc) $\mathcal{C}$ if

(1) the functions $x(t)$ and $z(t)$ are twice continuously differentiable
(2) for each $t \in I$, at least one of the derivatives $\dfrac{dx}{dt}, \dfrac{dz}{dt}$ is not zero
(3) for all points $\sigma, \tau \in I$, we have

$$(x(\sigma), z(\sigma)) \equiv (x(\tau), z(\tau))$$

if and only if $\sigma = \tau$.

1.1.1 Polar Coordinate System

The connection between the polar (r, φ) and Cartesian (x, z) coordinates is given by the formulas (Fig. 1.2)

Fig. 1.2 Polar coordinate system

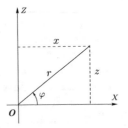

$$x = r \cos \varphi, \qquad z = r \sin \varphi$$
$$r \in \mathbb{R}^+, \qquad \varphi \in [0, 2\pi). \tag{1.1}$$

In the opposite direction we have

$$r = (x^2 + z^2)^{1/2}$$
$$\varphi = \arctan \frac{z}{x}. \tag{1.2}$$

1.1.2 Curvature

The curve $\mathcal{C}$ is referred to as a continuous if the defining functions are continuous. Respectively, it is called algebraic, if the equation(s) are algebraic. In all other cases, the curve is transcendental.

Usually we shall denote the arc-length of the curve $\mathcal{C}$ by s, and if we chose it to parameterize the curve, we will have

$$\frac{d\mathbf{x}}{ds} = \frac{dx}{ds}\mathbf{e}_1 + \frac{dz}{ds}\mathbf{e}_2, \tag{1.3}$$

where $\mathbf{e}_1$ and $\mathbf{e}_2$ are the unit vectors along the coordinate axes. For the scalar obtained by squaring the vector in (1.3), we can write

$$\frac{d\mathbf{x}}{ds} \cdot \frac{d\mathbf{x}}{ds} = \left(\frac{dx}{ds}\right)^2 + \left(\frac{dz}{ds}\right)^2. \tag{1.4}$$

Due to the fact that all our consideration, are in the Euclidean space, we have the equation

$$ds^2 = dx^2 + dz^2, \tag{1.5}$$

and in this way, we obtain

$$\frac{d\mathbf{x}}{ds} \cdot \frac{d\mathbf{x}}{ds} = 1. \tag{1.6}$$

The last equation tells us that the tangent vector

$$\mathbf{T}(s) = \frac{d\mathbf{x}}{ds} \tag{1.7}$$

is of unit length, and in this case, s is called a natural parameter.

The curvature of the curve $\mathcal{C}$ is defined as the rate of the change of the slope angle θ of the tangent vector $\mathbf{T}$ toward the abscissa, i.e.,

$$\kappa(s) = \frac{d\theta(s)}{ds}. \tag{1.8}$$

The curvature radius $\mathcal{R}$ is the reciprocal of the absolute value of the curvature, and therefore

$$\mathcal{R} = \frac{1}{|\kappa(s)|} = \left|\frac{ds}{d\theta}\right|. \tag{1.9}$$

Hence,

$$\kappa = \frac{d\theta}{ds} = \frac{d\theta}{dx}\frac{dx}{ds} \tag{1.10}$$

and if we choose for the parameter the abscissa x, i.e., $\mathbf{x}(x) = (x, z(x))$, we can consequently write the equalities

$$\frac{d\theta}{dx} = \frac{d}{dx}(\arctan z'(x)) = \frac{z''(x)}{1 + [z'(x)]^2} \tag{1.11}$$

$$\kappa(x) = \frac{z''(x)}{(1 + [z'(x)]^2)^{3/2}}, \tag{1.12}$$

where in the second one, it is taken into account that, according to (1.5),

$$ds = \left(1 + \left(\frac{dz}{dx}\right)^2\right)^{1/2} dx. \tag{1.13}$$

If we use an arbitrary parametric representation of the curve, the above formula can be easily converted into

$$\kappa(t) = \frac{\dot{x}(t)\ddot{z}(t) - \ddot{x}(t)\dot{z}(t)}{(\dot{x}^2(t) + \dot{z}^2(t))^{3/2}}. \tag{1.14}$$

If we work with the polar coordinates (1.2)

$$x(\phi) = r(\phi)\cos\phi, \qquad z(\phi) = r(\phi)\sin\phi, \tag{1.15}$$

we can use (1.14) and some elementary transformations to derive the formula

$$\kappa(\phi) = \frac{r^2 + 2\dot{r}^2 - r\ddot{r}}{(r^2 + \dot{r}^2)^{3/2}}, \tag{1.16}$$

in which

$$r = r(\phi), \qquad \dot{r} = \frac{dr(\phi)}{d\phi}, \qquad \text{and} \qquad \ddot{r} = \frac{d^2 r(\phi)}{d\phi^2}. \tag{1.17}$$

Finally, if the curve $\mathcal{C}$ is defined with its Cartesian coordinates x and z by the equation $F(x, z) = 0$, we have

$$\kappa(x, z) = \frac{F_{xx} F_z^2 - 2 F_{xz} F_x F_z + F_{zz} F_x^2}{(F_x^2 + F_z^2)^{3/2}} \bigg|_{F \equiv 0}. \tag{1.18}$$

1.1.3 Frenet-Serret Equations

According to Eqs. (1.6) and (1.7), we have $\mathbf{T} \cdot \mathbf{T} = 1$. Differentiation of this scalar product with respect to the natural parameter s produces

$$\frac{d}{ds}(\mathbf{T} \cdot \mathbf{T}) = 2\mathbf{T} \cdot \frac{d\mathbf{T}}{ds} = 2\mathbf{T} \cdot \mathbf{T}' = 0. \tag{1.19}$$

This result indicates that the curvature vector $\dfrac{d\mathbf{T}}{ds} = \mathbf{T}'$ is orthogonal to $\mathbf{T}$, and therefore coincides up to a numerical factor with $\|\mathbf{N}\| = 1$ (see Fig. 1.1). From the figure, it is also clear that

$$\mathbf{T} = \frac{dx}{ds}\mathbf{e_1} + \frac{dz}{ds}\mathbf{e_2} = \cos\theta \mathbf{e_1} + \sin\theta \mathbf{e_2}, \tag{1.20}$$

and therefore

$$\mathbf{T}' = -\sin\theta \frac{d\theta}{ds}\mathbf{e_1} + \cos\theta \frac{d\theta}{ds}\mathbf{e_2} = \frac{d\theta}{ds}(-\sin\theta \mathbf{e_1} + \cos\theta \mathbf{e_2}) = \kappa(s)\mathbf{N}. \tag{1.21}$$

By differentiation of the normal vector $\mathbf{N} = -\sin\theta \mathbf{e_1} + \cos\theta \mathbf{e_2}$, also gets

$$\mathbf{N}' = -\kappa(s)\mathbf{T}. \tag{1.22}$$

Taking Eqs. (1.21) and (1.22) together, we end up with the so-called Frenet-Serret system of equations

$$\begin{pmatrix} \mathbf{T} \\ \mathbf{N} \end{pmatrix}' = \begin{bmatrix} 0 & \kappa(s) \\ -\kappa(s) & 0 \end{bmatrix} \begin{pmatrix} \mathbf{T} \\ \mathbf{N} \end{pmatrix}. \tag{1.23}$$

It is clear that the vectors $\mathbf{T}$ and $\mathbf{N}$ constitute a frame (repére mobile) in the plane *XOZ* which is defined along the curve $\mathcal{C}$. According to the fundamental theorem in the geometry of plane curves, every such curve is defined by its curvature $\kappa(s)$ up to Euclidean motion in this plane.

This actually means that the curve can be described independently from the coordinate system. The curvature itself is defined from its intrinsic equation, like those of Whewell (1849) and Cesaro (1896). In the first case, this is an equation relating s and θ, and in the second one, an equation containing the curvature radius $\mathcal{R}$ and s. Due to the relation $\mathcal{R}d\theta = ds$, these intrinsic equations are related and any one of them can be easily transformed into another.

The Frenet-Serret formulas are quite useful for an alternative representation of the curvature of plane curves. According to the definitions given above,

$$\mathbf{T}(s) = \frac{d\mathbf{x}}{ds}, \qquad \mathbf{T}' = \mathbf{x}' = \kappa(s)\mathbf{N},$$

which means that

$$\mathbf{T}' \cdot \mathbf{T}' = \mathbf{x}'' \cdot \mathbf{x}'' = \kappa^2(s)\mathbf{N} \cdot \mathbf{N} = \kappa^2(s),$$

and therefore we have the formula

$$\kappa(s) = \sqrt{\mathbf{x}'' \cdot \mathbf{x}''} = (x_{ss}^2 + z_{ss}^2)^{1/2}. \tag{1.24}$$

1.2 Space Curves

1.2.1 Unit Tangent Vector

If $\mathbf{x} = \mathbf{x}(s)$ is the natural representation of the regular space curve $\mathcal{C}$, we shall use the derivative $\dfrac{d\mathbf{x}}{ds} = \mathbf{x}'(s)$ to find the tangent of the curve at an arbitrary point $\mathbf{x}(s)$.

This definition is inspired by our geometrical intuition according to which

$$\mathbf{x}'(s) = \lim_{\Delta s \to 0} \frac{\mathbf{x}(s + \Delta s) - \mathbf{x}(s)}{\Delta s},$$

where the vector $\dfrac{\mathbf{x}(s + \Delta s) - \mathbf{x}(s)}{\Delta s}$ is the secant of $\mathcal{C}$, as shown in Fig. 1.3.

Besides, the vector $\mathbf{x}'(s)$ is of unit length, as we know that for the natural parameterization, the equality

$$\|\mathbf{x}'(s)\| = \left\| \frac{d\mathbf{x}}{ds} \right\| = 1$$

is fulfilled automatically. If $\mathbf{x} = \mathbf{x}(\tilde{s})$ is any other natural representation of $\mathcal{C}$, we have

$$s = \pm\tilde{s} + \text{const} \quad \text{and} \quad \frac{d\mathbf{x}}{d\tilde{s}} = \frac{d\mathbf{x}}{ds}\frac{ds}{d\tilde{s}} = \pm\frac{d\mathbf{x}}{ds}.$$

Fig. 1.3 The tangent and the
secant

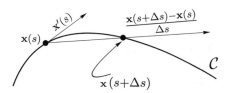

In other words, $\dfrac{d\mathbf{x}}{d\tilde{s}}$ has the same or opposite orientation as that of $\dfrac{d\mathbf{x}}{ds}$, depending
on the orientation of $\mathbf{x} = \mathbf{x}(\tilde{s})$. Consequently, $\mathbf{x}'$ is an oriented quantity. As can be
seen from Fig. 1.3, it has the same direction as that in which the parameter s increases.

The vector $\mathbf{x}'(s)$ is called the unit tangent vector of the oriented curve $\mathbf{x} = \mathbf{x}(s)$
at the point $\mathbf{x}(s)$, and we shall denote it with $\mathbf{T} = \mathbf{T}(s) = \mathbf{x}'(s)$. Besides the unit
tangent vector, we shall consider other geometrical quantities defined by the natural
parameter on the curve. Using the chain rule, i.e., the relation $\dfrac{ds}{dt} = \left\| \dfrac{d\mathbf{x}}{dt} \right\|$, we can
find these quantities using an arbitrary parameterization. For example, if $\mathbf{x} = \mathbf{x}(t)$ is
an arbitrary representation of the curve $\mathcal{C}$ with a fixed orientation of $\mathbf{x}(s)$, we have

$$\dot{\mathbf{x}}(t) = \frac{d\mathbf{x}}{dt} = \frac{d\mathbf{x}}{ds}\frac{ds}{dt} = \frac{ds}{dt}\frac{d\mathbf{x}}{ds} = \left\| \frac{d\mathbf{x}}{dt} \right\| \mathbf{T} = \|\dot{\mathbf{x}}\| \mathbf{T},$$

i.e., as expected, the derivative of the vector $\mathbf{x}$ has the same direction as $\mathbf{T}$, and
therefore is a tangent vector to the curve. That is why we have the formula

$$\mathbf{T} = \frac{\dot{\mathbf{x}}}{\|\dot{\mathbf{x}}\|}. \tag{1.25}$$

Example 1.1 Let us consider the helix $\mathbf{x} = a\cos t\mathbf{e}_1 + a\sin t\mathbf{e}_2 + ct\mathbf{e}_3$, where $a, c \neq 0$.
We have

$$\frac{d\mathbf{x}}{dt} = -a\sin t\mathbf{e}_1 + a\cos t\mathbf{e}_2 + c\mathbf{e}_3 \quad \text{and} \quad \left\| \frac{d\mathbf{x}}{dt} \right\| = \sqrt{a^2 + c^2}.$$

Therefore,

$$\mathbf{T} = \frac{d\mathbf{x}}{dt} \bigg/ \left\| \frac{d\mathbf{x}}{dt} \right\| = \frac{-a\sin t\mathbf{e}_1 + a\cos t\mathbf{e}_2 + c\mathbf{e}_3}{\sqrt{a^2 + c^2}}.$$

From the equation above, we have

$$\mathbf{T} \cdot \mathbf{e}_3 = \cos \sphericalangle(\mathbf{T}, \mathbf{e}_3) = \frac{c}{\sqrt{a^2 + c^2}} = \text{const},$$

i.e., the tangent vector toward the helix subtends a constant angle with the helix axis.

1.2.2   Tangent Line and Normal Plane

The line through any point $\mathbf{x}$ of a regular curve, parallel to the tangent vector at this point, is called a tangent of the curve C at the point $\mathbf{x}$. The tangent at the point $\overset{\circ}{\mathbf{x}}$ is given by the equation

$$\mathbf{x} = \overset{\circ}{\mathbf{x}} + \lambda\overset{\circ}{\mathbf{T}},$$

where $\overset{\circ}{\mathbf{T}} = \mathbf{T}(\overset{\circ}{t})$ is the unit tangent vector at the point $\overset{\circ}{\mathbf{x}}$.

The plane through the point $\mathbf{x}$, which is orthogonal to the tangent at $\mathbf{x}$, is called a normal plane of C at this point (see Fig. 1.4).

Consequently, the equation of the normal plane through the point $\overset{\circ}{\mathbf{x}}$ is

$$(\mathbf{x} - \overset{\circ}{\mathbf{x}}) \cdot \overset{\circ}{\mathbf{T}} = 0.$$

At this stage, it is convenient to introduce a second variable, say $\mathbf{z}$, by which we shall denote the points of other geometrical objects, connected to the curve $\mathbf{x}(t)$. For example, the equation of the tangent through an arbitrary point $\mathbf{x}$ on C can be written in the form

$$\mathbf{z} = \mathbf{x} + \lambda\mathbf{T}, \qquad -\infty < \lambda < \infty,$$

and that of the normal plane as

$$(\mathbf{z} - \mathbf{x}) \cdot \mathbf{T} = 0.$$

Also, using the fact that $\dot{\mathbf{x}}$ and $\mathbf{T}$ are parallel, we can present the above equations in the form

$$\mathbf{z} = \mathbf{x} + \lambda\dot{\mathbf{x}}, \qquad -\infty < \lambda < \infty$$

and

$$(\mathbf{z} - \mathbf{x}) \cdot \dot{\mathbf{x}} = 0.$$

Fig. 1.4   Normal plane

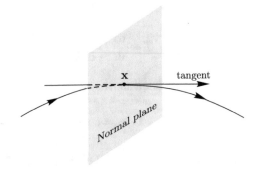

1.2.3 Curvature

Let us suppose that $\mathbf{x} = \mathbf{x}(s)$ is a regular curve of a class $m \geq 2$. Then, the tangent vector $\mathbf{T} = \mathbf{T}(s) = \mathbf{x}'(s)$ is at least of class C^1, and we can consider its derivative

$$\frac{d\mathbf{T}}{ds} = \mathbf{T}'(s) = \mathbf{x}''(s).$$

In this connection, let us notice that while the direction of the unit tangent vector depends on the orientation of C, the direction of $\mathbf{T}'$ does not depend on the chosen orientation. Actually, let $\mathbf{x} = \mathbf{x}(\tilde{s})$ is another natural representation of C with a unit tangent vector $\tilde{\mathbf{T}} = \dfrac{d\mathbf{x}}{d\tilde{s}}$.

Since $s = \pm\tilde{s} + \text{const}$, then

$$\frac{d\tilde{\mathbf{T}}}{d\tilde{s}} = \frac{d}{d\tilde{s}}\left(\frac{d\mathbf{x}}{d\tilde{s}}\right) = \frac{d}{d\tilde{s}}\left(\frac{d\mathbf{x}}{ds}\frac{ds}{d\tilde{s}}\right) = \pm\frac{d}{d\tilde{s}}\left(\frac{d\mathbf{x}}{ds}\right) = \pm\frac{d}{ds}\left(\frac{d\mathbf{x}}{ds}\right)\frac{ds}{d\tilde{s}}$$
$$= (\pm 1)^2 \frac{d}{ds}\left(\frac{d\mathbf{x}}{ds}\right) = \frac{d\mathbf{T}}{ds}.$$

The vector $\mathbf{T}'(s)$ is called a curvature vector of C at the point $\mathbf{x}(s)$, and we shall denote it with

$$\mathbb{k} = \mathbb{k}(s) = \mathbf{T}'(s).$$

As $\mathbf{T}$ is unit vector, the vector $\mathbb{k} = \mathbf{T}'$ is orthogonal to $\mathbf{T}$, and therefore parallel to the normal plane. When $\mathbb{k}$ is different from the zero vector, then its direction coincides with the direction in which the curve C is bending (Fig. 1.5).

By definition, the length of the curvature vector is $\|\mathbb{k}(s)\|$, which we shall denote with

$$|\kappa| = \|\mathbb{k}(s)\| \tag{1.26}$$

Fig. 1.5 Curvature vector

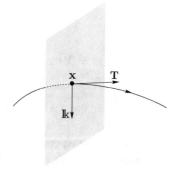

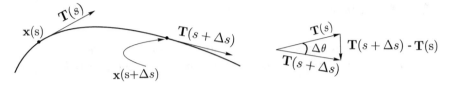

Fig. 1.6 The curvature and its geometrical interpretation

and refer to as a curvature of $\mathcal{C}$ at the point $\mathbf{x}(s)$. The reciprocal value of the curvature

$$\rho = \frac{1}{|\kappa|} = \frac{1}{\|\mathbf{k}(s)\|} \tag{1.27}$$

will be called a curvature radius of $\mathcal{C}$ at the point $\mathbf{x}(s)$. Every point of the curve $\mathcal{C}$, at which the curvature vector $\mathbf{k}$ becomes zero, is called an inflection point. At inflection points, the curvature κ is zero, and therefore the curvature radius ρ is infinite. A more geometrical description of the curvature will be of some interest as well. With this idea in mind, let us consider an arbitrary curve $\mathbf{x} = \mathbf{x}(s)$ of class $m \geq 2$, and let $\Delta\theta$ be the angle between the unit tangent vectors $\mathbf{T}(s)$ and $\mathbf{T}(s + \Delta s)$ at respective points $\mathbf{x}(s)$ and $\mathbf{x}(s + \Delta s)$.

The opposite side from the angle $\Delta\theta$ of the triangle in Fig. 1.6 is the base of an isosceles triangle, so that one can write the equalities

$$\|\mathbf{T}(s + \Delta s) - \mathbf{T}(s)\| = 2\sin\frac{\Delta\theta}{2} = \Delta\theta + o(\Delta\theta),$$

in which use has been made of elementary geometry and the explicit form of the Taylor series of the sine function. Then,

$$
\begin{aligned}
|\kappa| = \|\mathbf{T}'\| &= \left\| \lim_{\Delta s \to 0} \frac{\mathbf{T}(s + \delta s) - \mathbf{T}(s)}{\Delta s} \right\| \\
&= \lim_{\Delta s \to 0} \left\| \frac{\mathbf{T}(s + \delta s) - \mathbf{T}(s)}{\Delta s} \right\| = \lim_{\Delta s \to 0} \frac{\Delta\theta + o(\Delta\theta)}{\Delta s} \\
&= \lim_{\Delta s \to 0} \left[\frac{\Delta\theta}{\Delta s} \left(1 + \frac{o(\Delta\theta)}{\Delta\theta} \right) \right].
\end{aligned}
$$

As $\lim_{\Delta s \to 0} \Delta\theta = 0$, then $\lim_{\Delta s \to 0} \frac{o(\Delta\theta)}{\Delta\theta} = 0$, and therefore

$$|\kappa| = \lim_{\Delta s \to 0} \frac{\Delta\theta}{\Delta s} = \frac{d\theta}{ds}.$$

Example 1.2 Let us consider the circle $\mathbf{x} = a\cos t\mathbf{e}_1 + a\sin t\mathbf{e}_3$ in the *XOZ* plane with a radius a. In this case, we have

$$\frac{d\mathbf{x}}{dt} = -a\sin t\mathbf{e}_1 + a\cos t\mathbf{e}_3.$$

Then, $\left\|\dfrac{d\mathbf{x}}{dt}\right\| = a$, and therefore

$$\mathbf{T} = \frac{d\mathbf{x}}{dt} \Big/ \left\|\frac{d\mathbf{x}}{dt}\right\| = -\sin t\mathbf{e}_1 + \cos t\mathbf{e}_3$$

$$\mathbf{k} = \mathbf{T}' = \frac{d\mathbf{T}}{ds} = \frac{d\mathbf{T}}{dt}\frac{dt}{ds} = \frac{d\mathbf{T}}{dt}\frac{ds}{dt} = \frac{d\mathbf{T}}{dt} \Big/ \left\|\frac{d\mathbf{x}}{dt}\right\|$$

$$= -\frac{1}{a}(\cos t\mathbf{e}_1 + \sin t\mathbf{e}_3).$$

We shall note that $\mathbf{k}$ is directed to the origin of the coordinate system, and that the curvature in this case is constant, as $|\kappa| = \|\mathbf{k}\| = \dfrac{1}{a}$, and the curvature radius $\rho = \dfrac{1}{|\kappa|} = a$ coincides with the radius of the circle.

Example 1.3 For the helix $\mathbf{x} = \mathbf{x}(t) = a\cos t\mathbf{e}_1 + a\sin t\mathbf{e}_2 + ct\mathbf{e}_3, a > 0, c \neq 0$, we have

$$\frac{d\mathbf{x}}{dt} = -a\sin t\mathbf{e}_1 + a\cos t\mathbf{e}_2 + c\mathbf{e}_3, \qquad \left\|\frac{d\mathbf{x}}{dt}\right\| = \sqrt{a^2 + c^2}$$

$$\mathbf{T} = \frac{d\mathbf{x}}{dt} \Big/ \left\|\frac{d\mathbf{x}}{dt}\right\| = \frac{1}{\sqrt{a^2 + c^2}}(-a\sin t\mathbf{e}_1 + a\cos t\mathbf{e}_2 + c\mathbf{e}_3),$$

and therefore

$$\mathbf{k} = \mathbf{T}' = \frac{d\mathbf{T}}{dt} \Big/ \left\|\frac{d\mathbf{x}}{dt}\right\|$$

$$= -\frac{a}{\sqrt{a^2 + c^2}}(\cos t\mathbf{e}_1 + \sin t\mathbf{e}_2)/\sqrt{a^2 + c^2}$$

$$= -\frac{a}{a^2 + c^2}(\cos t\mathbf{e}_1 + \sin t\mathbf{e}_2).$$

Let us also notice that $\mathbf{k}$ is parallel to the *XOY* plane and directed toward to the *OZ* axis, as it is shown in Fig. 1.7. The curvature of the helix curve is constant and equal to

$$|\kappa| = \|\mathbf{k}\| = \frac{a}{a^2 + c^2}.$$

This situation suggests that we consider the case of curves $\mathcal{C}$, whose curvatures are equal to zero, i.e., $\|\mathbf{k}\| = 0$. This also means that $\|\mathbf{T}'\| = \mathbf{0}$. After the integration in the general case, we have $\mathbf{T} = \mathbf{a}, \mathbf{a} = \text{const} \neq \mathbf{0}$.

Since $\mathbf{T} = \mathbf{x}'$, after a new integration, we have

Fig. 1.7 A helix along the
OZ axis

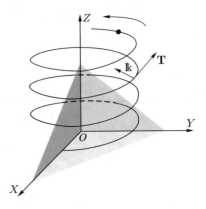

$$\mathbf{x} = s\,\mathbf{a} + \mathbf{c}, \qquad \mathbf{c} = \text{const},$$

and therefore $\mathcal{C}$ is a line through the point $\mathbf{c} = (c_1, c_2, c_3)$ which is parallel to the vector $\mathbf{a} = a_1\mathbf{e}_1 + a_2\mathbf{e}_2 + a_3\mathbf{e}_3$.

And vice versa, if $\mathcal{C}$ is the curve $\mathbf{x} = t\,\mathbf{a} + \mathbf{c}, \mathbf{a} \neq \mathbf{0}$, then

$$\mathbf{T} = \frac{d\mathbf{x}}{dt} \Big/ \left\| \frac{d\mathbf{x}}{dt} \right\| = \mathbf{a}/\|\mathbf{a}\| = \text{const} \quad \text{and} \quad \|\mathbb{k}\| = \|\mathbf{T}'\| = 0,$$

and therefore we can formulate

Theorem 1.1 *A regular curve of the class $m \geq 2$ is a line if and only if its curvature is equal to zero.*

Due to the fact that a curve is not always presented by its natural parameter and that in many cases passing to natural representation is connected with serious difficulties, it is useful to have a formula for the curvature for an arbitrary parameterization. For this purpose, we consider an arbitrary curve $\mathbf{x} = \mathbf{x}(t)$ of class $m \geq 2$. We have

$$\dot{\mathbf{x}} = \tfrac{d\mathbf{x}}{dt} = \tfrac{d\mathbf{x}}{ds}\tfrac{ds}{dt} = \dot{s}\mathbf{x}'$$

$$\ddot{\mathbf{x}} = \tfrac{d}{dt}(\dot{s}\mathbf{x}') = \ddot{s}\mathbf{x}' + \dot{s}\tfrac{d\mathbf{x}'}{dt} = \ddot{s}\mathbf{x}' + \dot{s}^2\mathbf{x}''$$

and

$$\dot{\mathbf{x}} \times \ddot{\mathbf{x}} = (\dot{s}\mathbf{x}') \times (\ddot{s}\mathbf{x}' + \dot{s}^2\mathbf{x}'')$$
$$= (\dot{s}\mathbf{x}') \times (\dot{s}^2\mathbf{x}'') = \dot{s}^3\mathbf{x}' \times \mathbf{x}'' = \|\dot{\mathbf{x}}\|^3(\mathbf{x}' \times \mathbf{x}'').$$

For writing down the last equation, we have used the fact that

$$\dot{s} = \frac{ds}{dt} = \|\dot{\mathbf{x}}\|.$$

Then,

$$\|\dot{\mathbf{x}} \times \ddot{\mathbf{x}}\| = \|\dot{\mathbf{x}}\|^3 \|\mathbf{x}' \times \mathbf{x}'\| = \|\dot{\mathbf{x}}\|^3 \|\mathbf{x}'\| \|\mathbf{x}''\| \sin \sphericalangle (\mathbf{x}', \mathbf{x}'').$$

But $\mathbf{x}' = \mathbf{T}$ and $\mathbf{x}'' = \mathbf{T}'$ are orthogonal, $\|\mathbf{x}'\| = 1$ and $\|\mathbf{x}''\| = \|\mathbf{T}'\| = \|\mathbb{k}\| = |\kappa|$. And finally,

$$|\kappa| = \frac{\|\dot{\mathbf{x}} \times \ddot{\mathbf{x}}\|}{\|\dot{\mathbf{x}}\|^3}. \tag{1.28}$$

1.2.4 Principal Normal Vector

The parallel to the curvature vector $\mathbb{k}$ of unit norm, chosen so that this vector is continuous on $\mathcal{C}$, will be denoted with $\mathbf{N}(s)$, and we shall call it the principal unit normal vector of the curve at the point $\mathbf{x}(s)$. In particular, if the curve $\mathcal{C}$ does not have inflection points, i.e., $\mathbb{k}(s) \neq 0$ for all s, we can take the vector

$$\mathbf{N}(s) = \mathbb{k}(s)/\|\mathbb{k}(s)\|,$$

which is just a unit vector with the direction of $\mathbb{k}(s)$. Obviously, for any straight line for which the curvature vector is zero, the principal unit normal vector is not defined. But when $\mathbf{N}(s)$ is given, the continuous function $\kappa(s)$ is fully determined by the equation

$$\mathbb{k}(s) = \kappa(s)\mathbf{N}(s). \tag{1.29}$$

At the points at which $\mathbf{N}$ has the same direction as $\mathbb{k}$, we have $\kappa(s) = \|\mathbb{k}(s)\|$, and when $\mathbf{N}$ has an opposite direction, $\kappa(s) = -\|\mathbb{k}(s)\|$. Finally, at the inflection points $\|\mathbb{k}(s)\| = 0$, $\kappa = 0$. The quantity $\kappa(s)$, defined in Eq. (1.29), is also called the curvature of $\mathcal{C}$ at the point $\mathbf{x}(s)$. However, and as far as the direction of $\mathbf{N}(s)$ is arbitrary, it should be noted that the function $\kappa(s)$ is defined up to a sign and that locally only its absolute value $\kappa = |\kappa|$, i.e., the curvature, defined above is the real characteristic of the curve.

If we evaluate the scalar product of both sides of Eq. (1.29) with $\mathbf{N}$ and take into account that $\mathbf{N} \cdot \mathbf{N} = \|\mathbf{N}\|^2 = 1$, we shall obtain

$$\kappa(s) = \mathbb{k}(s) \cdot \mathbf{N}(s). \tag{1.30}$$

1.2.5 Principal and Osculating Plane

The line through the point $\mathbf{x}$ on the curve $\mathcal{C}$, which is parallel to the principal normal of $\mathcal{C}$, is called the principal normal line at this point (Fig. 1.8).

Fig. 1.8 The normal and the
osculating planes

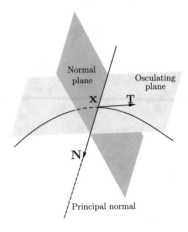

The plane which is parallel to the tangent and the principal normal is called an
osculating plane or contact plane for the curve $\mathcal{C}$ at the point $\mathbf{x}$. The first term was
introduced in 1686 by Leibniz, who chose the Latin word *osculare*, the translation
of which into English means *to kiss*. As the equation for the principal normal line is

$$\tilde{\mathbf{x}} = \mathbf{x} + \lambda\mathbf{N}, \qquad -\infty < \lambda < \infty,$$

for the equation of the contact plane, we have

$$[(\tilde{\mathbf{x}} - \mathbf{x})\mathbf{T}\mathbf{N}] = 0.$$

If we refer to the definition $\mathbf{T} = \mathbf{x}'$ and the fact that the vector $\mathbf{T}' = \mathbf{x}''$ is parallel to
$\mathbf{N}$, we can conclude that at the points for which $\lambda \neq 0$, the equation of the osculating
plane is given by the equation

$$[(\tilde{\mathbf{x}} - \mathbf{x})\mathbf{x}'\mathbf{x}''] = 0.$$

Here is a good place to recall that the tangent to the curve $\mathcal{C}$ at the given point was
defined as the limit position of the line through two neighboring points of the curve,
when the second one approaches the first. In this sense, the tangent is the line which
best approximates the curve at the chosen point. In the same way, the osculating
plane can be defined as the plane through three neighboring points of the curve when
the end ones approach the middle one. The tangent line and the osculating plane are
examples of geometrical objects, that have a contact of certain order with the curve.
From this comes the term 'contact plane'. The contact between curves and planes
will be studied further on, and the osculating plane will be reconsidered from this
point of view once more.

Example 1.4 Let us consider the helix

$$\mathbf{x} = \cos t\mathbf{e}_1 + \sin t\mathbf{e}_2 + t\mathbf{e}_3,$$

for which we have

$$\dot{\mathbf{x}} = -\sin t\mathbf{e}_1 + \cos t\mathbf{e}_2 + \mathbf{e}_3$$

$$\|\dot{\mathbf{x}}\| = \sqrt{\sin^2 t + \cos^2 t + 1} = \sqrt{2}$$

$$\mathbf{T} = \frac{\dot{\mathbf{x}}}{\|\dot{\mathbf{x}}\|} = \frac{1}{\sqrt{2}}(-\sin t\mathbf{e}_1 + \cos t\mathbf{e}_2 + \mathbf{e}_3)$$

$$\mathbb{k} = \mathbf{T}' = \frac{\dot{\mathbf{T}}}{\|\dot{\mathbf{x}}\|} = -\frac{1}{2}(\cos t\mathbf{e}_1 + \sin t\mathbf{e}_2),$$

and as $\mathbb{k} \neq \mathbf{0}$ for each t, one can write

$$\mathbf{N} = \frac{\mathbb{k}}{\|\mathbb{k}\|} = -(\cos t\mathbf{e}_1 + \sin t\mathbf{e}_2).$$

The equation of the principal normal at the point $t = \pi/2$ is

$$\tilde{\mathbf{x}} = \mathbf{x}(\pi/2) + \lambda\mathbf{N}(\pi/2) = (1-\lambda)\mathbf{e}_2 + (\pi/2)\mathbf{e}_3, \qquad -\infty < \lambda < \infty,$$

and that of the osculating plane through the same point is

$$\left[(\tilde{\mathbf{x}} - \mathbf{x}(\pi/2))\mathbf{TN}\right] = \det\begin{bmatrix} \tilde{x} & -\frac{1}{\sqrt{2}} & 0 \\ \tilde{y}-1 & 0 & -1 \\ \tilde{z}-\pi/2 & \frac{1}{\sqrt{2}} & 0 \end{bmatrix} = 0$$

or

$$\tilde{x} + \tilde{z} = \pi/2.$$

1.2.6 Binormal and Moving Frame

Let $\mathbf{x} = \mathbf{x}(s)$ be a regular curve $\mathcal{C}$ of class $m \geq 2$, on which $\mathbf{N}(s)$ is continuous. Then, at every point of $\mathcal{C}$, we have two orthogonal and continuous vectors—the unit tangent vector $\mathbf{T}$ and the unit principal normal vector $\mathbf{N}$. Let us consider the vector

$$\mathbf{B}(s) = \mathbf{T}(s) \times \mathbf{N}(s),$$

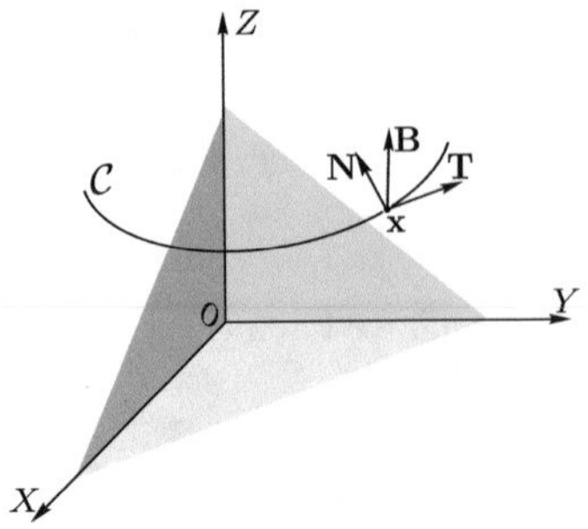

Fig. 1.9 The moving frame
associated with the curve $\mathcal{C}$

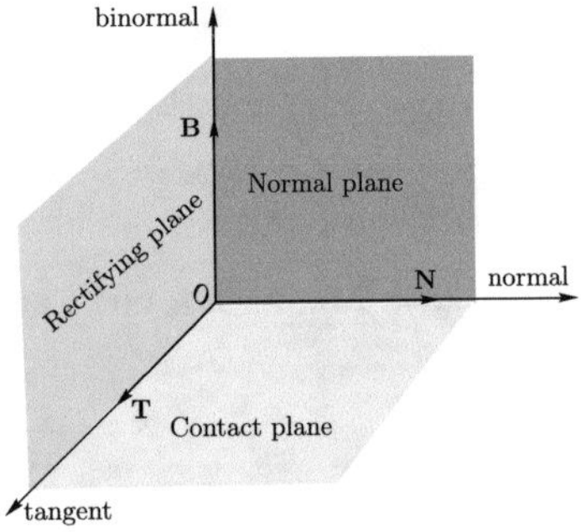

Fig. 1.10 Various planes
and lines associated with the
curve

which we notice is continuous and of unit length, i.e., $(\mathbf{T}, \mathbf{N}, \mathbf{B})$ form a right-handed
orthonormal basis. The vector $\mathbf{B}(s)$ is called a unit binormal vector of $\mathcal{C}$ at the point
$\mathbf{x}(s)$, and the triad $(\mathbf{T}(s), \mathbf{N}(s), \mathbf{B}(s))$ is called a moving frame of the curve $\mathcal{C}$.

The line through $\mathbf{x}(s)$ which is parallel to $\mathbf{B}$ is called the binormal of $\mathcal{C}$ at the point
$\mathbf{x}$, and the equation which describes it is (Fig. 1.9)

$$\tilde{\mathbf{x}} = \mathbf{x} + \lambda\mathbf{B}, \quad -\infty < \lambda < \infty.$$

The plane through a point $\mathbf{x}$ of $\mathcal{C}$ which is parallel to $\mathbf{T}$ and $\mathbf{B}$ is called a rectifying
plane at this point. The equation of the rectifying plane at the point $\mathbf{x}$ is

$$(\tilde{\mathbf{x}} - \mathbf{x}) \cdot \mathbf{N} = 0.$$

So at any points $\mathbf{x}$ of the curve $\mathcal{C}$, we have the following three special lines and planes
(see Fig. 1.10):

Tangent	$\tilde{\mathbf{x}} = \mathbf{x} + \lambda\mathbf{T}$
Normal	$\tilde{\mathbf{x}} = \mathbf{x} + \lambda\mathbf{N}$
Binormal	$\tilde{\mathbf{x}} = \mathbf{x} + \lambda\mathbf{B}$
Normal plane	$(\tilde{\mathbf{x}} - \mathbf{x}) \cdot \mathbf{T} = 0$
Rectifying plane	$(\tilde{\mathbf{x}} - \mathbf{x}) \cdot \mathbf{N} = 0$
Contact plane	$(\tilde{\mathbf{x}} - \mathbf{x}) \cdot \mathbf{B} = 0.$

Example 1.5 Let us consider the helix

$$\mathbf{x} = a \cos t \mathbf{e}_1 + a \sin t \mathbf{e}_2 + c t \mathbf{e}_3, \qquad a > 0, \qquad c \neq 0.$$

We have

$$\mathbf{T} = \frac{1}{\sqrt{a^2 + c^2}}(-a \sin t \mathbf{e}_1 + a \cos t \mathbf{e}_2 + c \mathbf{e}_3)$$

$$\mathbb{k} = -\frac{a}{\sqrt{a^2 + c^2}}(\cos t \mathbf{e}_1 + \sin t \mathbf{e}_2)$$

$$\mathbf{N} = \frac{\mathbb{k}}{\|\mathbb{k}\|} = -(\cos t \mathbf{e}_1 + \sin t \mathbf{e}_2)$$

$$\mathbf{B} = \mathbf{T} \times \mathbf{N} = \det \begin{bmatrix} \mathbf{e}_1 & -\frac{a}{\sqrt{a^2+c^2}}\sin t & -\cos t \\ \mathbf{e}_2 & \frac{a}{\sqrt{a^2+c^2}}\cos t & -\sin t \\ \mathbf{e}_3 & \frac{c}{\sqrt{a^2+c^2}} & 0 \end{bmatrix}$$

$$= \frac{1}{\sqrt{a^2 + c^2}}(c \sin t \mathbf{e}_1 - c \cos t \mathbf{e}_2 + a \mathbf{e}_3).$$

The equation of the rectifying plane at the point $\mathbf{x}(\mathring{t})$ is

$$(\tilde{\mathbf{x}} - \mathbf{x}(\mathring{t})) \cdot \mathbf{N}(\mathring{t}) = 0$$

or

$$(\tilde{x} - a \cos \mathring{t})(-\cos \mathring{t}) + (\tilde{y} - a \sin \mathring{t})(-\sin \mathring{t}) = 0,$$

and finally,

$$\tilde{x} \cos \mathring{t} + \tilde{y} \sin \mathring{t} = a.$$

Let us point out that all rectifying planes are parallel to the OZ axis.

1.2.7 Torsion

Let us now consider the regular curve $\mathbf{x} = \mathbf{x}(s)$ of class $m \geq 3$, along which the normal vector $\mathbf{N}$ is of class C^1. In this case, we can differentiate the defining equation $\mathbf{B}(s) = \mathbf{T}(s) \times \mathbf{N}(s)$, which gives us

$$\begin{aligned} \mathbf{B}'(s) &= \mathbf{T}'(s) \times \mathbf{N}(s) + \mathbf{T}(s) \times \mathbf{N}'(s) \\ &= \kappa(s)(\mathbf{N}(s) \times \mathbf{N}(s)) + \mathbf{T}(s) \times \mathbf{N}'(s) \\ &= \mathbf{T}(s) \times \mathbf{N}'(s), \end{aligned} \qquad (1.31)$$

where we have used (1.29), and the fact that $\mathbf{a} \times \mathbf{a} = 0$ for every $\mathbf{a}$. Since $\mathbf{N}(s)$ is a unit vector, $\mathbf{N}'(s)$ is orthogonal to $\mathbf{N}$, i.e., parallel to the rectifying plane, and therefore it can be presented as a linear combination of the vectors $\mathbf{T}$ and $\mathbf{B}$

$$\mathbf{N}'(s) = \alpha(s)\mathbf{T}(s) + \tau(s)\mathbf{B}(s).$$

Inserting this relation into (1.31), we have

$$\begin{aligned} \mathbf{B}'(s) &= \mathbf{T}(s) \times \mathbf{N}'(s) = \mathbf{T}(s) \times (\alpha(s)\mathbf{T}(s) + \tau(s)\mathbf{B}(s)) \\ &= \tau(s)\mathbf{T}(s) \times \mathbf{B}(s) = -\tau(s)\mathbf{B}(s) \times \mathbf{T}(s) \end{aligned}$$

or

$$\mathbf{B}'(s) = -\tau(s)\mathbf{N}(s), \qquad (1.32)$$

where we have used the fact that $(\mathbf{T}, \mathbf{N}, \mathbf{B})$ is a right-hand-oriented orthonormal basis. If we take the scalar product of (1.32) with $\mathbf{N}$, we have the formula

$$\tau(s) = -\mathbf{N}(s) \cdot \mathbf{B}'(s).$$

The continuous function $\tau(s)$, defined from formula (1.32), is called the torsion of $\mathcal{C}$ at the point $\mathbf{x}(s)$. Instead of torsion, some older books use the term 'second curvature of the curve'.

We shall note that the sign of τ does not depend on the direction of $\mathbf{N}$ and the orientation of $\mathcal{C}$, and therefore is an intrinsic property of the curve (although it depends on its position in $\mathbb{R}^3$). Let us prove this statement. For that purpose, we assume that we have chosen $\tilde{\mathbf{N}} = -\mathbf{N}(s)$ instead of $\mathbf{N}$, and consequently

$$\tilde{\mathbf{B}} = \mathbf{T} \times \tilde{\mathbf{N}} = \mathbf{T} \times (-\mathbf{N}) = -\mathbf{T} \times \mathbf{N} = -\mathbf{B}.$$

But then,

$$\tilde{\tau} = -\tilde{\mathbf{N}} \cdot \tilde{\mathbf{B}}' = -(-\mathbf{N}) \cdot (-\mathbf{B}') = -\mathbf{N} \cdot \mathbf{B}' = \tau,$$

i.e., we have proved that τ does not depend on the direction of $\mathbf{N}$. Now let us also change the orientation of the curve, which, as we know, means that $s = -\tilde{s} + \text{const.}$ Then, $\tilde{\mathbf{T}} = -\mathbf{T}$ and

$$\tilde{\mathbf{B}} = \tilde{\mathbf{T}} \times \mathbf{N} = -\mathbf{T} \times \mathbf{N} = -\mathbf{B}$$

and

$$\frac{d\tilde{\mathbf{B}}}{d\tilde{s}} = \frac{d\tilde{\mathbf{B}}}{ds}\frac{ds}{d\tilde{s}} = \frac{ds}{d\tilde{s}}\frac{d\tilde{\mathbf{B}}}{ds} = (-1)\left(-\frac{d\mathbf{B}}{ds}\right) = \frac{d\mathbf{B}}{ds},$$

which means at the end that

$$\tilde{\tau} = -\tilde{\mathbf{N}}\cdot\frac{d\tilde{\mathbf{B}}}{d\tilde{s}} = -\mathbf{N}\cdot\frac{d\mathbf{B}}{ds} = \tau,$$

as required.

Example 1.6 Let us consider again the helix

$$\mathbf{x} = a\cos t\mathbf{e}_1 + a\sin t\mathbf{e}_2 + ct\mathbf{e}_3, \qquad a > 0, \quad c \neq 0.$$

According to *Example* 1.5, we have

$$\mathbf{B}(t) = \frac{c}{\sqrt{a^2 + c^2}}(\sin t\mathbf{e}_1 - \cos t\mathbf{e}_2),$$

and therefore

$$\mathbf{B}'(s) = \frac{d\mathbf{B}}{ds} = \frac{d\mathbf{B}}{dt}\bigg/\left\|\frac{d\mathbf{x}}{dt}\right\| = \frac{c}{a^2 + c^2}(\cos t\mathbf{e}_1 + \sin t\mathbf{e}_2).$$

For the torsion of the helix, we have

$$\tau = -\mathbf{N}\cdot\mathbf{B}' = -\frac{c}{a^2 + c^2}(-(\cos t\mathbf{e}_1 + \sin t\mathbf{e}_2))\cdot(\cos t\mathbf{e}_1 + \sin t\mathbf{e}_2) = \frac{c}{a^2 + c^2}.$$

In this case, the torsion is constant, and one can also notice that if $c > 0$ (i.e., $\tau > 0$), the curve is a right-hand-oriented helix, as shown in Fig. 1.11a. If $c < 0$ (i.e., $\tau < 0$), the curve is a left-hand-oriented helix of the type, illustrated in Fig. 1.11b.

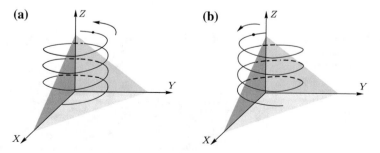

Fig. 1.11 The *right* and *left hand* oriented helices

As the sign of τ is intrinsic property, we can conclude that the two curves cannot be imposed upon each other.

If the torsion is identically zero on the curve $\mathbf{x} = \mathbf{x}(s)$, i.e., if $\tau(s) \equiv 0$, then $\mathbf{B}'(s) = -\phi(s)\mathbf{N}(s) \equiv \mathbf{0}$, and therefore $\mathbf{B} = \text{const} = \mathring{\mathbf{B}}$.

Let us now consider

$$\frac{d}{ds}(\mathring{\mathbf{B}} . \mathbf{x}) = \mathring{\mathbf{B}} . \mathbf{x}' = \mathring{\mathbf{B}} . \mathbf{T}.$$

Taking into account that $\mathbf{T}$ and $\mathring{\mathbf{B}}$ are orthogonal, one can conclude that $\dfrac{d}{ds}(\mathring{\mathbf{B}} . \mathbf{x}) = 0$, and by integrating this equation, we have

$$\mathring{\mathbf{B}} . \mathbf{x} = \text{const},$$

which makes obvious that $\mathbf{x} = \mathbf{x}(s)$ is a plane curve, lying in the plane $\mathring{\mathbf{B}} . \mathbf{x} = \text{const}$.

In particular, $\mathbf{x} = \mathbf{x}(s)$ lies in its contact plane, as is shown in Fig. 1.12. The opposite statement is also true, and we can formulate

Theorem 1.2 *A curve of class $m \geq 3$ along which $\mathbf{N}$ is of class C^1, is a plane curve if and only if its torsion is identically zero.*

From now aside from those cases in which we explicitly mention something else, we shall suppose that the curves are of class $m \geq 3$, and that along them, $\mathbf{N}$ is of class C^1. In this case, τ is a continuous function and κ, $\mathbf{T}$, $\mathbf{N}$ and $\mathbf{B}$ are functions of class C^1.

The torsion of the space curve was defined assuming that it is parameterized by its natural parameter. The corresponding formula for arbitrary parameterization is

$$\tau(t) = \frac{[\dot{\mathbf{x}}\ddot{\mathbf{x}}\dddot{\mathbf{x}}]}{\|\dot{\mathbf{x}} \times \ddot{\mathbf{x}}\|^2}. \tag{1.33}$$

It is valid under the condition that the curvature κ at the point does not vanish.

Let us go to the proof. We are assuming that $\mathbf{x} = \mathbf{x}(t)$ is some regular curve and that we are at a point on it for which $\kappa(t) \neq 0$. In this case, we have

Fig. 1.12 Plane curve (its binormal coincides with the unit normal vector to the plane)

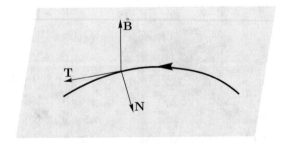

$$\mathbf{x}' = \frac{d\mathbf{x}}{ds} = \frac{d\mathbf{x}}{dt}\frac{dt}{ds} = \dot{\mathbf{x}}t'$$

$$\mathbf{x}'' = \frac{d}{ds}(\dot{\mathbf{x}}t') = \frac{d}{ds}(\dot{\mathbf{x}})t' + \dot{\mathbf{x}}t'' = \ddot{\mathbf{x}}t'^2 + \dot{\mathbf{x}}t''$$

$$\mathbf{x}''' = \frac{d}{ds}(\ddot{\mathbf{x}}t'^2 + \dot{\mathbf{x}}t'') = \dddot{\mathbf{x}}t'^3 + 2\ddot{\mathbf{x}}t't'' + \ddot{\mathbf{x}}t't'' + \dot{\mathbf{x}}t'''$$

$$= \dddot{\mathbf{x}}t'^3 + 3\ddot{\mathbf{x}}t't'' + \dot{\mathbf{x}}t'''.$$

Therefore,

$$
\begin{aligned}
[\mathbf{x}'\mathbf{x}''\mathbf{x}'''] &= [(\dot{\mathbf{x}}t')(\ddot{\mathbf{x}}t'^2 + \dot{\mathbf{x}}t'')(\dddot{\mathbf{x}}t'^3 + 3\ddot{\mathbf{x}}t't'' + \dot{\mathbf{x}}t''')] \\
&= (\dot{\mathbf{x}}t') \cdot ((\ddot{\mathbf{x}} \times \dddot{\mathbf{x}})t'^5 + (\ddot{\mathbf{x}} \times \dot{\mathbf{x}})t'^2 t''' \\
&\quad + (\dot{\mathbf{x}} \times \dddot{\mathbf{x}})t'^3 t'' + (\dot{\mathbf{x}} \times \ddot{\mathbf{x}})3t't'^2) \\
&= [\dot{\mathbf{x}}\ddot{\mathbf{x}}\dddot{\mathbf{x}}]t'^6,
\end{aligned}
\tag{1.34}
$$

as all other terms when expanding the triple product of vectors are identically zero. Before continuing with the proof, we take advantage of the fact that $\mathbf{T}$, $\mathbf{N}$ and $\mathbf{B}$ form a frame, and therefore every vector of E^3 can be expressed in terms of them. By definition,

$$\mathbf{x}'' = \mathbf{T}' = \kappa\mathbf{N}, \tag{1.35}$$

and after differentiation, we have

$$
\begin{aligned}
\mathbf{x}''' &= \kappa'\mathbf{N} + \kappa\mathbf{N}' = \kappa'\mathbf{N} + \kappa(\mathbf{B} \times \mathbf{T})' \\
&= \kappa'\mathbf{N} + \kappa\mathbf{B}' \times \mathbf{T} + \kappa\mathbf{B} \times \mathbf{T}' \\
&= \kappa'\mathbf{N} + \kappa(-\tau\mathbf{N}) \times \mathbf{T} + \kappa\mathbf{B} \times (\kappa\mathbf{N}) \\
&= \kappa'\mathbf{N} + \kappa\tau\mathbf{T} \times \mathbf{N} + \kappa^2\mathbf{B} \times \mathbf{N} \\
&= \kappa'\mathbf{N} + \kappa\tau\mathbf{B} - \kappa^2\mathbf{T},
\end{aligned}
$$

where we have utilized the fact that $\mathbf{T}' = \kappa\mathbf{N}$, $\mathbf{B}' = -\tau\mathbf{N}$, $\mathbf{N} = \mathbf{B} \times \mathbf{T}$, $\mathbf{B} = \mathbf{T} \times \mathbf{N}$ and $\mathbf{T} = \mathbf{N} \times \mathbf{B}$.

In a completely analogous way, we can calculate

$$
\begin{aligned}
\mathbf{x}'' \times \mathbf{x}''' &= \kappa\mathbf{N} \times (\kappa'\mathbf{N} + \kappa\tau\mathbf{B} - \kappa^2\mathbf{T}) = \kappa^2\tau\mathbf{N} \times \mathbf{B} - \kappa^3\mathbf{N} \times \mathbf{T} \\
&= \kappa^2\tau\mathbf{T} + \kappa^3\mathbf{B},
\end{aligned}
$$

and finally,

$$
\begin{aligned}
\mathbf{x}'\mathbf{x}''\mathbf{x}'''t &= \mathbf{x}' \cdot (\mathbf{x}'' \times \mathbf{x}''') = \mathbf{T} \cdot (\kappa^2\tau\mathbf{T} + \kappa^3\mathbf{B}) = \kappa^2\tau\mathbf{T} \cdot \mathbf{T} \\
&= \kappa^2\tau,
\end{aligned}
\tag{1.36}
$$

because $\mathbf{T} \cdot \mathbf{B} = 0$ and $\mathbf{T} \cdot \mathbf{T} = 1$. Now we can come back to the torsion formula, writing the results from (1.34) and (1.36) in the form

$$[\mathbf{x}'\mathbf{x}''\mathbf{x}'''] = \frac{[\dot{\mathbf{x}}\ddot{\mathbf{x}}\dddot{\mathbf{x}}]}{\|\dot{\mathbf{x}}\|^6}$$

and

$$\tau = \frac{[\mathbf{x}'\mathbf{x}''\mathbf{x}''']}{\kappa^2}.$$

Using the results above and the formula for the curvature (1.29)

$$\kappa = \frac{\|\dot{\mathbf{x}} \times \ddot{\mathbf{x}}\|}{\|\dot{\mathbf{x}}\|^3},$$

we have, finally,

$$\tau = \frac{[\dot{\mathbf{x}}\ddot{\mathbf{x}}\dddot{\mathbf{x}}]}{\kappa^2 \|\dot{\mathbf{x}}\|^6} = \frac{[\dot{\mathbf{x}}\ddot{\mathbf{x}}\dddot{\mathbf{x}}]}{\|\dot{\mathbf{x}} \times \ddot{\mathbf{x}}\|^2}.$$

1.2.8 Frenet-Serret Equations of Space Curves

Theorem 1.3 *The vectors* $\mathbf{T}$, $\mathbf{N}$ *and* $\mathbf{B}$ *on the space curve* $\mathbf{x} = \mathbf{x}(s)$ *satisfy the following relations, which are known also, as the* Frenet-Serret *formulas:*

$$\begin{aligned}
\mathbf{T}' &= & \kappa\mathbf{N} & \\
\mathbf{N}' &= -\kappa\mathbf{T} &+ & \tau\mathbf{B} \\
\mathbf{B}' &= & -\tau\mathbf{N}. &
\end{aligned} \qquad (1.37)$$

Proof Since the first and third equations are already proven (see (1.32) and (1.35)), we continue with proof of the second equation. Towards this aim, we shall take advantage of the fact that $\mathbf{T}$, $\mathbf{N}$ and $\mathbf{B}$ form an orthonormal basis (i.e., $\mathbf{N} = \mathbf{B} \times \mathbf{T}$) and the two other equations. We have

$$\begin{aligned}
\mathbf{N}' = (\mathbf{B} \times \mathbf{T})' &= \mathbf{B}' \times \mathbf{T} + \mathbf{B} \times \mathbf{T}' \\
&= (-\tau\mathbf{N}) \times \mathbf{T} + \mathbf{B} \times (\kappa\mathbf{N}) \\
&= -\tau\mathbf{N} \times \mathbf{T} + \kappa\mathbf{B} \times \mathbf{N} \\
&= \tau\mathbf{T} \times \mathbf{N} - \kappa\mathbf{N} \times \mathbf{B} \\
&= \tau\mathbf{B} - \kappa\mathbf{T} = -\kappa\mathbf{T} + \tau\mathbf{B}.
\end{aligned}$$

If we introduce the matrices

$$R(s) = \begin{bmatrix} T_1(s) & T_2(s) & T_3(s) \\ N_1(s) & N_2(s) & N_3(s) \\ B_1(s) & B_2(s) & B_3(s) \end{bmatrix}, \qquad \Omega(s) = \begin{bmatrix} 0 & \kappa(s) & 0 \\ -\kappa(s) & 0 & \tau(s) \\ 0 & -\tau(s) & 0 \end{bmatrix}$$

and

$$R'(s) = \frac{dR}{ds} = \begin{bmatrix} \frac{dT_1}{ds} & \frac{dT_2}{ds} & \frac{dT_3}{ds} \\ \frac{dN_1}{ds} & \frac{dN_2}{ds} & \frac{dN_3}{ds} \\ \frac{dB_1}{ds} & \frac{dB_2}{ds} & \frac{dB_3}{ds} \end{bmatrix},$$

we can write the Frenet-Serret equations in the form

$$\frac{dR}{ds} = \Omega R, \tag{1.38}$$

which, besides being quite compact, has some other advantages which we shall further use.

For example, as **T**, **N** and **B** form an orthonormal basis, we can conclude that the matrix $R(s)$ is an orthogonal matrix, i.e.,

$$R^t(s)R(s) = I, \tag{1.39}$$

where R^t denotes the transpose of R and I is the three-dimensional identity matrix. Another useful observation is that the matrix Ω is antisymmetric, i.e.,

$$\Omega^t = -\Omega. \tag{1.40}$$

To prove that (1.40) is fulfilled at every point of the curve, we differentiate the left-hand side of (1.39), and in this way obtain

$$\frac{d}{ds}(R^t R) = \frac{dR^t}{ds}R + R^t\frac{dR}{ds} = \left(\frac{dR}{ds}\right)^t R + R^t\frac{dR}{ds}$$
$$= (\Omega R)^t R + R^t \Omega R = R^t \Omega^t R + R^t \Omega R$$
$$= R^t(\Omega^t + \Omega)R = 0,$$

where we have utilized the fact that for arbitrary matrices A and B, the identities $(AB)' = A'B + AB'$, $(A')^t = (A^t)'$, $(AB)^t = B^t A^t$ are always satisfied, as well as Eq. (1.38) and the fact that $R^t R$ is the identity matrix along the curve.

1.2.9 Main Theorem in the Local Theory of Curves

The most important consequence from the Frenet-Serret equations is the fundamental theorem for existence and uniqueness of space curves, namely,

Theorem 1.4 *Let $\kappa(s)$ and $\tau(s)$ be arbitrary continuous functions on the interval* $a \leq s \leq c$. *Then, up to its position in space, there exists one and only one space curve C which has as its curvature and torsion, respectively, $\kappa(s)$ and $\tau(s)$, and s is the natural parameter on the curve.*

Proof We shall start proving the uniqueness assuming that through a fixed point $\overset{\circ}{\mathbf{x}} = \mathbf{x}(\overset{\circ}{s})$ with given $\kappa(s)$ and $\tau(s)$ pass two different curves C and $\tilde{C}$, with frames $(\mathbf{T}, \mathbf{N}, \mathbf{B})$ and $(\tilde{\mathbf{T}}, \tilde{\mathbf{N}}, \tilde{\mathbf{B}})$ for which the respective Frenet-Serret equations are true, i.e.,

$$
\begin{aligned}
\mathbf{T}' &= \kappa \mathbf{N} & \tilde{\mathbf{T}}' &= \kappa \tilde{\mathbf{N}} \\
\mathbf{N}' &= -\kappa \mathbf{T} + \tau \mathbf{B} & \tilde{\mathbf{N}}' &= -\kappa \tilde{\mathbf{T}} + \tau \tilde{\mathbf{B}} \\
\mathbf{B}' &= -\tau \mathbf{N} & \tilde{\mathbf{B}}' &= -\tau \tilde{\mathbf{N}}
\end{aligned}
$$

with some initial conditions $\mathbf{T}(\overset{\circ}{s}) \equiv \tilde{\mathbf{T}}(\overset{\circ}{s})$, $\mathbf{N}(\overset{\circ}{s}) \equiv \tilde{\mathbf{N}}(\overset{\circ}{s})$ and $\mathbf{B}(\overset{\circ}{s}) \equiv \tilde{\mathbf{B}}(\overset{\circ}{s})$. Let us introduce, as a measure of deviation of the frames, the quantity

$$
\Delta(R, \tilde{R}) = \frac{1}{2}(\|\mathbf{T} - \tilde{\mathbf{T}}\|^2 + \|\mathbf{N} - \tilde{\mathbf{N}}\|^2 + \|\mathbf{B} - \tilde{\mathbf{B}}\|^2)
$$

and differentiate this function. We have

$$
\begin{aligned}
\frac{\mathrm{d}}{\mathrm{d}s} \Delta(R, \tilde{R}) &= \frac{1}{2} \frac{\mathrm{d}}{\mathrm{d}s}(\|\mathbf{T} - \tilde{\mathbf{T}}\|^2 + \|\mathbf{N} - \tilde{\mathbf{N}}\|^2 + \|\mathbf{B} - \tilde{\mathbf{B}}\|^2) \\
&= (\mathbf{T} - \tilde{\mathbf{T}}) . (\mathbf{T}' - \tilde{\mathbf{T}}') + (\mathbf{N} - \tilde{\mathbf{N}}) . (\mathbf{N}' - \tilde{\mathbf{N}}') \\
&\quad + (\mathbf{B} - \tilde{\mathbf{B}}) . (\mathbf{B}' - \tilde{\mathbf{B}}') \\
&= \kappa(\mathbf{T} - \tilde{\mathbf{T}}) . (\mathbf{N} - \tilde{\mathbf{N}}) - \kappa(\mathbf{N} - \tilde{\mathbf{N}}) . (\mathbf{T} - \tilde{\mathbf{T}}) \\
&\quad + \tau(\mathbf{N} - \tilde{\mathbf{N}}) . (\mathbf{B} - \tilde{\mathbf{B}}) - \tau(\mathbf{B} - \tilde{\mathbf{B}}) . (\mathbf{N} - \tilde{\mathbf{N}}) = 0.
\end{aligned}
$$

This means that the function $\Delta(R, \tilde{R})$ is constant for every s within the interval. By its very definition, it is zero at the point $\overset{\circ}{s}$, and therefore identically zero on the whole interval I by continuity.

On the other hand, $\Delta(R, \tilde{R})$ is a sum of three positive summands, and as a consequence of this, each of them should itself be equal to zero, i.e., $\mathbf{T} \equiv \tilde{\mathbf{T}}$, $\mathbf{N} \equiv \tilde{\mathbf{N}}$ and $\mathbf{B} \equiv \tilde{\mathbf{B}}$ for every $s \in I$. With this, the uniqueness of the curve is proven. The proof for its existence relies on the respective in the theory of Ordinary Differential Equations (with given initial conditions) about the solutions to the first order system of linear differential equations (1.38).

Before going to the proof, we will take advantage of the statement of the theorem in order to introduce the following:

Definition 1.1 The equations $\kappa = \kappa(s)$ and $\tau = \tau(s)$, which give the curvature and the torsion of the curve as functions of the arclength s, are called natural, essential or intrinsic equations of the curve, because according to the above-formulated theorem, they completely define the curve.

In the general case, the solutions to the Frenet-Serret equations cannot be obtained through direct integration, but there is a method introduced by Sophus Lie for reducing the system of equations (1.37) to an equations of first order, known as Riccati equations. The latter has been very well studied, and more detail can be found in the books of Eisenhart (1909) and Struik (1988).

1.3 Surfaces

1.3.1 Surfaces and Parametric Curves

Every surface S in the Cartesian coordinate system specified by the orthonormal vectors $\mathbf{e}_1$, $\mathbf{e}_2$ and $\mathbf{e}_3$ is given by three functions of two real parameters u and v, namely,

$$x = x(u, v), \qquad y = y(u, v), \qquad z = z(u, v). \tag{1.41}$$

Here, x, y and z are considered to be single-valued and continuous functions of u and v that are well defined in $U \subset \mathbb{R}^2$. The parameters u and v are called curvilinear coordinates of the surface. When we fix one of these parameters and leave the other to change, we get a curve, which is called a parametric curve of the surface, as shown in Fig. 1.13. For many purposes, it is useful to write these three Eqs. (1.41) as one vector equation, i.e.,

$$\mathbf{x}(u, v) = x(u, v)\mathbf{e}_1 + y(u, v)\mathbf{e}_2 + z(u, v)\mathbf{e}_3. \tag{1.42}$$

The differential increment $\mathbf{dx}$ associated with the translation of the point M into the infinitesimally near point $\tilde{M}$ on the surface S can be written in the form

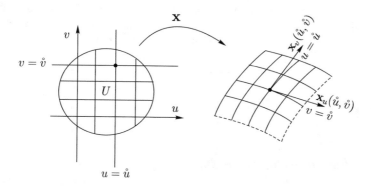

Fig. 1.13 Parametric representation and coordinate lines

$$\mathbf{dx} = \mathbf{x}_u du + \mathbf{x}_v dv, \tag{1.43}$$

where we have introduced the notations

$$\mathbf{x}_u = \frac{\partial \mathbf{x}}{\partial u}, \qquad \mathbf{x}_v = \frac{\partial \mathbf{x}}{\partial v} \tag{1.44}$$

for the partial derivatives of the vector $\mathbf{x}$. Every point at which they are not zero is called regular.

1.3.2 First Fundamental Form

The square of the length of the translational vector $\mathbf{dx}$ can be evaluated by taking the scalar product of $\mathbf{dx}$ with itself, i.e.,

$$ds^2 = \mathbf{dx}.\mathbf{dx} = E du^2 + 2F du dv + G dv^2, \tag{1.45}$$

where the following standard notations have been used:

$$E = \mathbf{x}_u.\mathbf{x}_u, \qquad F = \mathbf{x}_u.\mathbf{x}_v, \qquad G = \mathbf{x}_v.\mathbf{x}_v. \tag{1.46}$$

The right-hand side of (1.45) is known as the first fundamental form of the surface S defined by the vector $\mathbf{x}(u, v)$, and E, F and G are the so-called coefficients of the first fundamental form. Along the parametric curves, the differential arclength takes the form

$$\begin{aligned} ds_u &= \sqrt{E} du \quad \text{along the curve with the constant } v \\ ds_v &= \sqrt{G} dv \quad \text{along the curve with constant } u. \end{aligned} \tag{1.47}$$

Here, we note that while $\mathbf{x}_u$ and $\mathbf{x}_v$ are tangent to the curves with constant v, respectively u, the quantity F becomes zero only if the parametric curves generate orthogonal net. In such cases, it is convenient to write the first fundamental form as

$$ds^2 = A_u^2 du^2 + A_v^2 dv^2, \tag{1.48}$$

where $A_u = \sqrt{E}$, $A_v = \sqrt{G}$ and $F \equiv 0$.

1.3.3 Normal Vector to the Surface

At every regular point M from the surface, there is a normal vector $\mathbf{n}(u, v)$ which is perpendicular to $\mathbf{x}_u$ and $\mathbf{x}_v$, and therefore to the tangent plane at M, which contains these vectors. The unit normal vector is parallel to the vector product of tangent vectors $\mathbf{x}_u$ and $\mathbf{x}_v$. Hence, we get the unit vector when we divide the vector by its length. So, for $\mathbf{n}(u, v)$, we have

$$\mathbf{n}(u, v) = \frac{\mathbf{x}_u \times \mathbf{x}_v}{|\mathbf{x}_u \times \mathbf{x}_v|}. \tag{1.49}$$

From the vector algebra, we also have the relations

$$|\mathbf{x}_u \times \mathbf{x}_v| = |\mathbf{x}_u||\mathbf{x}_v| \sin \psi \tag{1.50}$$

and

$$\mathbf{x}_u.\mathbf{x}_v = |\mathbf{x}_u||\mathbf{x}_v| \cos \psi, \tag{1.51}$$

where ψ is the angle between the vectors $\mathbf{x}_u$ and $\mathbf{x}_v$. From the above equation and (1.46), we have

$$\cos \psi = F/\sqrt{EG}, \tag{1.52}$$

and therefore

$$\sin \psi = \sqrt{(EG - F^2)/EG}. \tag{1.53}$$

The final expression for the unit normal vector is

$$\mathbf{n}(u, v) = \frac{\mathbf{x}_u \times \mathbf{x}_v}{\mathcal{E}}, \qquad \mathcal{E} = \sqrt{EG - F^2}, \tag{1.54}$$

under the condition that $\mathcal{E}$ does not become zero. We have to note that the principal normal $\mathbf{N}$ of the curve on the surface is not always orthogonal to the surface (i.e., in the general case $\mathbf{N}.\mathbf{n} \neq 1$).

As in the case of the main normals to the curves, the direction of the normal to the surface is arbitrary. That is why we can assume that the parametric curves are always organized in the way that the normal $\mathbf{n}$ of $\mathcal{S}$ is directed from the concave to the convex side of the surface.

1.3.4 A Second Fundamental Form

Let us consider a curve on a surface and use the properties of the curvature vector to calculate another important characteristic of the surface called the second fundamental form. To do this, we shall recall that the curvature vector $\mathbf{k}$ is given by $d\mathbf{T}/ds$ where $\mathbf{T}$ is the unit tangent vector to the curve. Since we can decompose every vector on $\mathcal{S}$ in components which are orthogonal and tangential to the surface, we have

$$\mathbf{k} = \frac{d\mathbf{T}}{ds} = \mathbf{k}_n + \mathbf{k}_t, \tag{1.55}$$

where $\mathbf{k}_n$ and $\mathbf{k}_t$ denote, respectively, the normal and the tangential components of the curvature vector. Here, we are interested mainly in the first one. Since $\mathbf{k}_n$ lies on the normal to the surface, it is proportional to $\mathbf{n}$ and can be expressed in the form

$$\mathbf{k}_n = -\kappa_n \mathbf{n}. \tag{1.56}$$

The minus sign means that the direction of the curvature vector $\mathbf{k}$ is opposite to that of the normal vector $\mathbf{n}$. Since $\mathbf{n}$ is perpendicular to $\mathbf{T}$, the differentiation of the scalar product $\mathbf{n}.\mathbf{T} = 0$ with respect to s along the curve of the surface gives,

$$\frac{d\mathbf{n}}{ds} \cdot \mathbf{T} = -\mathbf{n} \cdot \frac{d\mathbf{T}}{ds}. \tag{1.57}$$

If we take the scalar product of Eq. (1.56) with $\mathbf{n}$, we will obtain

$$\mathbf{k}_n \cdot \mathbf{n} = -\kappa_n, \tag{1.58}$$

and the scalar product of Eq. (1.55) with $\mathbf{n}$ will produce (since $\mathbf{n}$ is perpendicular to $\mathbf{k}_t$)

$$\mathbf{n} \cdot \frac{d\mathbf{T}}{ds} = \mathbf{n}. \mathbf{k}_n. \tag{1.59}$$

Finally, if we combine Eqs. (1.59), (1.58) and (1.57), we will have, as a result,

$$\kappa_n = -\frac{d\mathbf{x}.d\mathbf{n}}{d\mathbf{x}.d\mathbf{x}}, \tag{1.60}$$

where we have used that $ds^2 = d\mathbf{x}.d\mathbf{x}$. If we remember that

$$d\mathbf{x} = \mathbf{x}_u du + \mathbf{x}_v dv, \qquad d\mathbf{n} = \mathbf{n}_u du + \mathbf{n}_v dv \tag{1.61}$$

and put these expressions in Eq. (1.60), we will find that

$$\kappa_n = \frac{L du^2 + 2M du dv + N dv^2}{E du^2 + 2F du dv + G dv^2} = \frac{II}{I}. \tag{1.62}$$

Here, L, M and N denote the coefficients of the second fundamental form $II = L du^2 + 2M du dv + N dv^2$ of the surface $\mathcal{S}$, which are defined as follows:

$$L = -\mathbf{x}_u.\mathbf{n}_u, \qquad 2M = -(\mathbf{x}_u.\mathbf{n}_v + \mathbf{x}_v.\mathbf{n}_u), \qquad N = -\mathbf{x}_v.\mathbf{n}_v. \tag{1.63}$$

Differentiating the equations $\mathbf{x}_u.\mathbf{n} = 0$ and $\mathbf{x}_v.\mathbf{n} = 0$ leads to alternative expressions for the coefficients of the second fundamental form, which in many cases are convenient from a computational point of view:

$$L = \mathbf{x}_{uu}.\mathbf{n}, \quad M = \mathbf{x}_{uv}.\mathbf{n}, \quad N = \mathbf{x}_{vv}.\mathbf{n}. \tag{1.64}$$

The coefficients of the first fundamental form E, F and G are in one to one correspondence with the metric g of the surface

$$g = \begin{bmatrix} E & F \\ F & G \end{bmatrix}, \tag{1.65}$$

and as such describe the stretching which is necessary to cover the surface with a piece of the parametric plane. The coefficients of the second fundamental form L, M and N have a connection with acceleration, and therefore with curvature. They can also be associated with another 2×2 matrix which will be denoted by

$$h = \begin{bmatrix} L & M \\ M & N \end{bmatrix}. \tag{1.66}$$

There are some classical formulas which describe two types of curvatures at every point of the surface. These are the so-called Gauss and mean (average) curvatures, whose standard notations are K and H. These formulas are

$$K = \frac{LN - M^2}{EG - F^2}, \quad H = \frac{EN + GL - 2FM}{2(EG - F^2)} \tag{1.67}$$

and can be expressed via the invariants of the matrix $g^{-1}h$, i.e.,

$$K = \det(g^{-1}h), \quad 2H = \operatorname{trace}(g^{-1}h). \tag{1.68}$$

The above matrices are quite useful for the compact recording of different differentiation formulas like

$$\frac{\partial \mathbf{n}}{\partial u} = -\sum_{k=1}^{2} (g^{-1}h)_{1k}\mathbf{x}_k, \quad \frac{\partial \mathbf{n}}{\partial v} = -\sum_{k=1}^{2} (g^{-1}h)_{2k}\mathbf{x}_k, \quad k = 1, 2 = u, v, \tag{1.69}$$

which are known as Weingarten formulas (see Oprea 2000).

Since the triad $\mathbf{x}_u$, $\mathbf{x}_v$ and $\mathbf{n}$ forms a moving frame in space, the vectors $\mathbf{x}_{uu}$, $\mathbf{x}_{uv}$ and $\mathbf{x}_w$ can be decomposed in the form

$$\begin{aligned} \mathbf{x}_{uu} &= \Gamma_{11}^1 \mathbf{x}_u + \Gamma_{11}^2 \mathbf{x}_v + \lambda\mathbf{n} \\ \mathbf{x}_{uv} &= \Gamma_{12}^1 \mathbf{x}_u + \Gamma_{12}^2 \mathbf{x}_v + \mu\mathbf{n} \\ \mathbf{x}_{vv} &= \Gamma_{22}^1 \mathbf{x}_u + \Gamma_{22}^2 \mathbf{x}_v + \nu\mathbf{n}. \end{aligned} \tag{1.70}$$

The six coefficients Γ_{ij}^k, $i, j, k = 1, 2$ are called Christoffel symbols, and it can be proven (see any book on differential geometry), that they are determined by the equations

$$\Gamma_{11}^1 = \frac{GE_u - 2FF_u + FE_v}{2(EG - F^2)}, \qquad\qquad \Gamma_{11}^2 = \frac{2EF_u - EE_v - FE_u}{2(EG - F^2)}$$

$$\Gamma_{12}^1 = \frac{GE_v - FG_u}{2(EG - F^2)}, \qquad\qquad \Gamma_{12}^2 = \frac{EG_u - FE_v}{2(EG - F^2)} \qquad (1.71)$$

$$\Gamma_{22}^1 = \frac{2GF_v - GG_u - FG_v}{2(EG - F^2)}, \qquad\qquad \Gamma_{22}^2 = \frac{EG_v - 2FF_v + FG_u}{2(EG - F^2)}.$$

The other three coefficients λ, μ and ν, coincide with the coefficients of the second fundamental form L, M and N.

It should be noted that the Christoffel symbols depend only on the coefficients of the first fundamental form and their first derivatives, and therefore characterize the intrinsic geometry of the surface. A useful addition to the Gauss equations are the equations which express the derivatives $\mathbf{n}_u$ and $\mathbf{n}_v$ through the basic vectors of the moving trihedron $\mathbf{x}_u$, $\mathbf{x}_v$ and $\mathbf{n}$. Since $\mathbf{n}_u$ and $\mathbf{n}_v$ lie in the tangent plane, we can write

$$\mathbf{n}_u = p_1 \mathbf{x}_u + q_1 \mathbf{x}_v, \qquad \mathbf{n}_v = p_2 \mathbf{x}_u + q_2 \mathbf{x}_v, \qquad (1.72)$$

where p_1, p_2, q_1 and q_2 are some (unknown) coefficients. To find them, we multiply Eqs. (1.72) consecutively with $\mathbf{x}_u$ and $\mathbf{x}_v$. In that way, we have

$$\begin{aligned}
\mathbf{n}_u.\mathbf{x}_u = -L &= p_1 \mathbf{x}_u.\mathbf{x}_u + q_1 \mathbf{x}_v.\mathbf{x}_u = p_1 E + q_1 F \\
\mathbf{n}_u.\mathbf{x}_v = -M &= p_1 \mathbf{x}_u.\mathbf{x}_v + q_1 \mathbf{x}_v.\mathbf{x}_v = p_1 F + q_1 G \\
\mathbf{n}_v.\mathbf{x}_u = -M &= p_2 \mathbf{x}_v.\mathbf{x}_u + q_2 \mathbf{x}_v.\mathbf{x}_u = p_2 E + q_2 F \\
\mathbf{n}_v.\mathbf{x}_v = -N &= p_2 \mathbf{x}_u.\mathbf{x}_v + q_2 \mathbf{x}_v.\mathbf{x}_v = p_2 F + q_2 G.
\end{aligned} \qquad (1.73)$$

From the first two equations in (1.73), we obtain

$$p_1 = \frac{MF - LG}{EG - F^2}, \qquad q_1 = \frac{LF - ME}{EG - F^2},$$

and from the next two,

$$p_2 = \frac{NF - MG}{EG - F^2}, \qquad q_2 = \frac{MF - NE}{EG - F^2}.$$

This allows us to rewrite Eqs. (1.72) as

$$\begin{aligned}
\mathbf{n}_u &= \frac{MF - LG}{EG - F^2}\mathbf{x}_u + \frac{LF - ME}{EG - F^2}\mathbf{x}_v \\
\mathbf{n}_v &= \frac{NF - MG}{EG - F^2}\mathbf{x}_u + \frac{MF - NE}{EG - F^2}\mathbf{x}_v.
\end{aligned} \qquad (1.74)$$

In this form, they are known as the Weingarten equations. One has to note the distinction between the Christoffel symbols and the coefficients p_1, p_2, q_1 and q_2 in the second group of equations for the derivatives (1.72), which depend on the coefficients of both the first and the second fundamental forms.

1.3.5 Integrability Conditions

The Gauss equations (1.70) determine the surface coordinates $\mathbf{x} = \mathbf{x}[u, v]$ as functions of u and v in the terms of differential equations. These equations are not independent, and that is why some conditions for integrability have to be satisfied. Due to the fact that the coefficients of these equations have to be at least once differentiable, and they themselves are expressed by the derivatives of the metric, it is obvious that it is reasonable to assume at least triple differentiation of the vector function $\mathbf{x}[u, v]$ and that the eight third order partial derivatives $\mathbf{x}_{uuu}, \mathbf{x}_{uuv}, \mathbf{x}_{uvu}, \mathbf{x}_{uvv}, \mathbf{x}_{vuu}, \mathbf{x}_{vuv}, \mathbf{x}_{vvu}$ and $\mathbf{x}_{vvv}$ are continuous functions of u and v in their domain of definition Ω. Again, not all of them are independent, and we have the relations

$$\mathbf{x}_{uuv} = \mathbf{x}_{uvu} = \mathbf{x}_{vuu}$$
$$\mathbf{x}_{uvv} = \mathbf{x}_{vuv} = \mathbf{x}_{vvu},$$

which can be written in the form

$$(\mathbf{x}_{uu})_v = (\mathbf{x}_{uv})_u = (\mathbf{x}_{vu})_u$$
$$(\mathbf{x}_{uv})_v = (\mathbf{x}_{vu})_v = (\mathbf{x}_{vv})_u.$$

Let us consider the equation $(\mathbf{x}_{uu})_v = (\mathbf{x}_{uv})_u$, which we can rewrite as

$$(\mathbf{x}_{uu})_v - (\mathbf{x}_{uv})_u = \mathbf{0}. \tag{1.75}$$

If we use the Gauss equations (1.70), we can write the Eq. (1.75) in the form

$$
\begin{aligned}
\mathbf{0} &= \frac{\partial}{\partial v}(\Gamma_{11}^1 \mathbf{x}_u + \Gamma_{11}^2 \mathbf{x}_v + L\mathbf{n}) - \frac{\partial}{\partial u}(\Gamma_{12}^1 \mathbf{x}_u + \Gamma_{12}^2 \mathbf{x}_v + M\mathbf{n}) \\
&= \left(\frac{\partial \Gamma_{11}^1}{\partial v} - \frac{\partial \Gamma_{12}^1}{\partial u}\right)\mathbf{x}_u + \left(\frac{\partial \Gamma_{11}^2}{\partial v} - \frac{\partial \Gamma_{12}^2}{\partial u}\right)\mathbf{x}_v + \left(\frac{\partial L}{\partial v} - \frac{\partial M}{\partial u}\right)\mathbf{n} \\
&\quad - \Gamma_{12}^1 \mathbf{x}_{uu} + (\Gamma_{11}^1 - \Gamma_{12}^2)\mathbf{x}_{uv} + \Gamma_{11}^2 \mathbf{x}_{vv} - M\mathbf{n}_u + L\mathbf{n}_v.
\end{aligned} \tag{1.76}
$$

If we decompose the vectors $\mathbf{x}_{uu}, \mathbf{x}_{uv}$ and $\mathbf{x}_{vv}$ in (1.76), using the Gauss equations (1.70) and those of Weingarten (1.74) for $\mathbf{n}_u$ and $\mathbf{n}_v$, and rearrange the so-obtained terms, we will have

$$\mathbf{0} = \left(\frac{\partial \Gamma_{11}^1}{\partial v} - \frac{\partial \Gamma_{12}^1}{\partial u} - \Gamma_{12}^1 \Gamma_{12}^2 + \Gamma_{22}^1 \Gamma_{11}^1 + L\frac{NF-MG}{EG-F^2} - M\frac{MF-LG}{EG-F^2} \right) \mathbf{x}_u$$

$$+ \left(\frac{\partial \Gamma_{11}^1}{\partial v} - \frac{\partial \Gamma_{12}^2}{\partial u} - \Gamma_{11}^1 \Gamma_{12}^2 + \Gamma_{11}^2 \Gamma_{22}^2 - \Gamma_{12}^1 \Gamma_{11}^1 - \Gamma_{12}^2 \Gamma_{12}^2 + L\frac{MF-NE}{EG-F^2} \right.$$

$$\left. - M\frac{LF-ME}{EG-F^2} \right) \mathbf{x}_v + (L_v - M_u - L\Gamma_{12}^1 + M(\Gamma_{11}^1 - \Gamma_{12}^2) + N\Gamma_{11}^2)\mathbf{n}$$

$$= \left(\frac{\partial \Gamma_{11}^1}{\partial v} - \frac{\partial \Gamma_{12}^1}{\partial u} - \Gamma_{12}^1 \Gamma_{12}^2 + \Gamma_{22}^1 \Gamma_{11}^1 + F\frac{LN-M^2}{EG-F^2} \right) \mathbf{x}_u$$

$$+ \left(\frac{\partial \Gamma_{11}^2}{\partial v} - \frac{\partial \Gamma_{12}^2}{\partial u} - \Gamma_{11}^1 \Gamma_{12}^2 + \Gamma_{11}^2 \Gamma_{22}^2 - \Gamma_{12}^1 \Gamma_{11}^1 - \Gamma_{12}^2 \Gamma_{12}^2 - E\frac{LN-M^2}{EG-F^2} \right) \mathbf{x}_v$$

$$+ (L_v - M_u - L\Gamma_{12}^1 + M(\Gamma_{11}^1 - \Gamma_{12}^2) + N\Gamma_{11}^2)\mathbf{n}. \tag{1.77}$$

If we take into account that the vectors $\mathbf{x}_u$, $\mathbf{x}_v$ and $\mathbf{n}$ are linearly independent, we can conclude that the expressions in front of them are equal to zero identically. The so-obtained equations are known as Gauss-Codazzi-Minardi (G-C-M) compatibility equations. Proceeding in absolutely the same way but starting with $\mathbf{x}_{uvv} = \mathbf{x}_{vvu}$, we end up with another three equations of the same type. When the coefficients in front of $\mathbf{x}_u$ and $\mathbf{x}_v$ are equated to zero, we obtain the equations

$$FK = \frac{\partial \Gamma_{12}^1}{\partial u} - \frac{\partial \Gamma_{11}^1}{\partial v} + \Gamma_{12}^1 \Gamma_{12}^2 - \Gamma_{22}^1 \Gamma_{11}^2$$

$$EK = \frac{\partial \Gamma_{11}^2}{\partial v} - \frac{\partial \Gamma_{12}^2}{\partial u} + \Gamma_{11}^1 \Gamma_{12}^2 + \Gamma_{11}^2 \Gamma_{22}^2 - \Gamma_{12}^1 \Gamma_{11}^2 - \Gamma_{12}^2 \Gamma_{12}^2$$

$$GK = \frac{\partial \Gamma_{22}^1}{\partial u} - \frac{\partial \Gamma_{12}^1}{\partial v} + \Gamma_{11}^1 \Gamma_{22}^1 + \Gamma_{12}^1 \Gamma_{22}^2 - \Gamma_{12}^1 \Gamma_{11}^1 + \Gamma_{22}^1 \Gamma_{12}^2 \tag{1.78}$$

$$FK = \frac{\partial \Gamma_{12}^2}{\partial v} - \frac{\partial \Gamma_{22}^2}{\partial u} + \Gamma_{22}^1 \Gamma_{11}^2 - \Gamma_{12}^1 \Gamma_{1}^2,$$

which are the essence of the most important theorem of the 19th century, known as the Theorema Egregium (the most beautiful theorem), as coined by Gauss himself.

Theorem 1.5 (Gauss 1828, see McClearly 1994, p.148) *The Gauss curvature of the surface S of class $m \geq 3$ in E^3 is a function only of the coefficients of the first fundamental form and their derivatives.*

To ensure that the Gauss theorem is true, it is enough to put the Christoffel symbols (1.71) in any of Eqs. (1.78). We shall also note that the last two are equivalent to each other. When the normal components (i.e., the coefficients in front of $\mathbf{n}$) become zero, the following two equations are completely independent and are known in the literature as Codazzi-Mainardi-Peterson (C-M-P) equations. Their explicit form is

$$L_v - M_u = L\Gamma_{12}^1 - M(\Gamma_{11}^1 - \Gamma_{12}^2) - N\Gamma_{11}^2$$

$$M_v - N_u = L\Gamma_{22}^1 - M(\Gamma_{12}^1 - \Gamma_{22}^2) - N\Gamma_{12}^2. \tag{1.79}$$

We note that the compatibility conditions, generated by the Weingarten equations (1.74) and in correspondence with the fact that $\mathbf{n}_{uv}$ and $\mathbf{n}_{vu}$ are equal, coincide with the C-M-P equations (1.79).

If $\mathbf{t}$ is a unit tangent vector at the point M of the surface $\mathcal{S}$, then we can find the normal curvature $\kappa_n(\mathbf{t})$ in the direction of $\mathbf{t}$ by cutting $\mathcal{S}$ with a plane defined by $\mathbf{t}$ and the normal $\mathbf{n}$ and take the curvature (at the point M) of the curve of intersection. In some sense, this is the most fundamental type of curvature that can be associated with a given surface. The so-described procedure of its generation defines a continuous function $\kappa_n \colon S^1 \to \mathbb{R}$ (in this case, we identify the set of the unit vectors in $\mathbb{R}^2$ with the circle S^1). Since S^1 is a compact manifold, the function κ_n reaches its maximum κ_1 and minimum κ_2 values. These values are called the principal curvatures, and one can prove that $K = \kappa_1\kappa_2$ and $H = (k_1 + k_2)/2$ (see Oprea 2007). From here, we can easily obtain the relations

$$\kappa_1 = H + \sqrt{H^2 - K}, \qquad \kappa_2 = H - \sqrt{H^2 - K}. \tag{1.80}$$

Since further on, the main object in our considerations will be the rotational surfaces whose general parameterizations (up to a permutation of the Cartesian coordinates x, y and z) are given by the formulas

$$\mathbf{x}(u, v) = (h(u)\cos(v),\ h(u)\sin(v),\ g(u)), \tag{1.81}$$

we can use the above formulas to find their curvatures. At the beginning, we have to notice that in this case, we have $F = 0 = M$, and that is why the general formulas (1.67) simplify considerably. Respectively, the formulas for the principal curvatures reduce to the expressions

$$k_\mu = \frac{g''h' - g'h''}{(g'^2 + h'^2)^{3/2}}, \qquad k_\pi = \frac{g'}{h\sqrt{g'^2 + h'^2}}. \tag{1.82}$$

The indices μ and π in them refer to the directions of meridians $(h(u), g(u))$ and parallel circles on the surface.

1.4 Variational Calculus

1.4.1 Euler-Lagrange Equation

One of the main tasks in variational calculus is finding the extremum of the integral J below, in which the integrand F (often call the Lagrangian) is a function of the independent variable x and the dependent variables $z(x)$ and

$$\dot{z} = \frac{dz(x)}{dx},$$

i.e.,

$$F = F(x, z(x), \dot{z}(x)).$$

If the integration end points, a and c are fixed, then F defines the functional J by the formula

$$J = \int_a^c F(x, z, \dot{z})dx. \tag{1.83}$$

The type of the function $z(x)$ is unknown and should be determined by the condition that J has an extremum for all possible variations $\tilde{z}(x)$ of $z(x)$, for which the points $(a, z(a))$ and $(c, z(c))$ stay fixed. We shall note that from the formulation of the problem itself, it is clear that this task is more complicated than that of finding the extremum of a given function.

From Fig. 1.14, it is clear that we can assign a mechanical meaning to the function $z(x)$ and its arbitrary variation $\tilde{z}(x)$ can be viewed as the trajectories in the plane XOZ passing through the points $A = (a, z(a))$ and $C = (c, z(c))$.

In order to find the optimal trajectory, it is enough to introduce the real parameter ε and the arbitrary function $\tau(x)$, which becomes identically zero at the endpoints, i.e.,

$$\tau(a) = \tau(c) = 0, \tag{1.84}$$

by which we can describe all possible trajectories having the property that they connect the chosen points A and C. Every trajectory of the above type can be written in the form

$$\tilde{z}(x, \varepsilon) = z(x) + \varepsilon\tau(x). \tag{1.85}$$

Fig. 1.14 The extremal trajectory $z(x)$ is drawn as a *bold line* and its variation as a *dotted one*

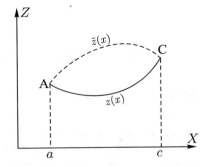

If we put (1.85) in (1.83), then the value of J will depend on the trajectory via the infinitesimal parameter ε, and this dependence is defined by the equality

$$J(\varepsilon) = \int_a^c F(x, \tilde{z}(x, \varepsilon), \tilde{z}_x(x, \varepsilon)) dx. \tag{1.86}$$

The requirement that $J(\varepsilon)$ has to be extremal implies the equation

$$\frac{\partial J(\varepsilon)}{\partial \varepsilon} \bigg|_{\varepsilon=0} = 0. \tag{1.87}$$

When evaluating the integral (1.86), we shall assume that the functions $\tilde{z}(x, \varepsilon)$ and $\tilde{z}_x(x, \varepsilon)$ are continuous, and therefore we can differentiate under the integral sign. Differentiating both sides of (1.86) with respect to ε, we have

$$\frac{\partial J(\varepsilon)}{\partial \varepsilon} = \int_a^c \left(\frac{\partial F}{\partial \tilde{z}} \frac{\partial \tilde{z}}{\partial \varepsilon} + \frac{\partial F}{\partial \tilde{z}_x} \frac{\partial \tilde{z}_x}{\partial \varepsilon} \right) dx, \tag{1.88}$$

while from (1.85), it follows that

$$\frac{\partial \tilde{z}(x, \varepsilon)}{\partial \varepsilon} = \tau(x), \qquad \frac{\partial \tilde{z}_x(x, \varepsilon)}{\partial \varepsilon} = \frac{d\tau(x)}{dx}.$$

Then,

$$\frac{\partial J(\varepsilon)}{\partial \varepsilon} = \int_a^c \left(\frac{\partial F}{\partial \tilde{z}} \tau(x) + \frac{\partial F}{\partial \tilde{z}_x} \frac{\partial \tau}{\partial x} \right) dx. \tag{1.89}$$

The second term under the integral in (1.89) can be integrated by parts, and in this way, we obtain

$$\int_a^c \frac{\partial F}{\partial \tilde{z}_x} \frac{\partial \tau}{\partial x} dx = \frac{\partial F}{\partial \tilde{z}_x} \tau(x) \bigg|_a^c - \int_a^c \tau(x) \frac{d}{dx} \frac{\partial F}{\partial \tilde{z}_x} dx. \tag{1.90}$$

Let us however take into account that according to (1.84), the first term on the right-hand side (1.90) becomes zero at the end points, and therefore

$$\int_a^c \frac{\partial F}{\partial \tilde{z}_x} \frac{\partial \tau}{\partial x} dx = - \int_a^c \tau(x) \frac{d}{dx} \frac{\partial F}{\partial \tilde{z}_x} dx. \tag{1.91}$$

Putting (1.91) back into (1.88), we have

$$\frac{\partial J(\varepsilon)}{\partial \varepsilon} = \int_a^c \left[\frac{\partial F}{\partial \tilde{z}} - \frac{d}{dx} \frac{\partial F}{\partial \tilde{z}_x} \right] \tau(x) dx. \tag{1.92}$$

The extremum of $J(\varepsilon)$ is attained when $\dfrac{\partial J(\varepsilon)}{\partial \varepsilon} = 0$, and we have chosen this to happens when $\varepsilon \equiv 0$. In this case, from (1.92), we can conclude that

$$\frac{\partial J(\varepsilon)}{\partial \varepsilon}\bigg|_{\varepsilon=0} = 0 = \int_a^c \left[\frac{\partial F}{\partial z} - \frac{\mathrm{d}}{\mathrm{d}x}\frac{\partial F}{\partial z_x}\right]\tau(x)\mathrm{d}x, \tag{1.93}$$

where $z(x) = \tilde{z}(x, 0)$ is the extremal trajectory. Hence, (1.93) is valid for an arbitrary function $\tau(x)$ satisfying the boundary conditions (1.84), which ultimately means that

$$\frac{\partial F}{\partial z} - \frac{\mathrm{d}}{\mathrm{d}x}\frac{\partial F}{\partial z_x} = 0. \tag{1.94}$$

Equation (1.94), which in the general case is an equation of second order, is known in the literature as the Euler-Lagrange equation.

1.4.2 First Integrals of the Euler-Lagrange Equation

Further on, we shall consider some specific cases in which the Euler-Lagrange equation can be integrated relatively easily and we shall give some examples.

One such case is when the Lagrangian does not contain the dependant variable z, as the Euler-Lagrange equation then simplifies considerably, namely,

$$\frac{\mathrm{d}}{\mathrm{d}x}\frac{\partial F}{\partial z_x} = 0. \tag{1.95}$$

From the above, we immediately have

$$\frac{\partial F}{\partial z_x} = C, \tag{1.96}$$

where C is arbitrarily constant. In this case, the Euler-Lagrange equation reduces to an equation which contains only x and z_x, and therefore it is of first order. If, in addition, the Lagrangian F does not contain the independent variable x, the partial derivative $\dfrac{\partial F}{\partial z_x}$ is a function only of z_x, and therefore the solutions to Eq. (1.96) are of the form

$$z_x = \text{const.} \tag{1.97}$$

It follows that the solutions to the initial problem for finding the extremals of J when the Lagrangian F depends only on z_x are necessarily linear functions of x.

Example 1.7 Let us consider the problem of finding the curve of minimal length which connects two arbitrary points in the plane XOZ.

Fig. 1.15 The straight segment is extremum curve connecting two points in the plane

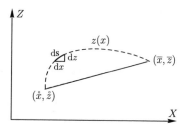

If these points have Cartesian coordinates $(\mathring{x}, \mathring{z})$ and $(\overline{x}, \overline{z})$, the infinitesimal distance along the varied curve $z(x)$ is given by the equation

$$ds = (dx^2 + dz^2)^{\frac{1}{2}}. \tag{1.98}$$

The total length of the curve connecting $(\mathring{x}, \mathring{z})$ and $(\overline{x}, \overline{z})$ is

$$s = \int_{\mathring{x}}^{\overline{x}} (dx^2 + dz^2)^{\frac{1}{2}} = \int_{\mathring{x}}^{\overline{x}} \left(1 + \left(\frac{dz}{dx}\right)^2 \right)^{\frac{1}{2}} dx. \tag{1.99}$$

We are interested in the extremum (minimum) of s. The integrand $F = (1 + z_x^2)^{\frac{1}{2}}$ is exactly of the type we have just considered, and the respective Euler-Lagrange equation in this particular case reduces to

$$\frac{d}{dx} \frac{\partial F}{\partial z_x} = 0.$$

It can be integrated once, giving

$$\frac{\partial F}{\partial z_x} = \frac{z_x}{(1 + z_x^2)^{\frac{1}{2}}} = C, \tag{1.100}$$

where C is an integration constant. By (1.100), one can conclude immediately that z_x is also a constant which we will denote by a. Integrating once more, we have

$$z(x) = ax + c,$$

and in this way, we prove that the desired trajectory is just a line in the plane. The constants a and c can be determined by the linear system of equations

$$a\mathring{x} + c = \mathring{z}, \qquad a\overline{x} + c = \overline{z}, \tag{1.101}$$

which is obtained from the requirement that the curve (in this case, a line) has to pass through the starting and end point, respectively.

Proof that we have a real minimum requires further analysis, but we shall refer in this case to physical intuition (see Fig. 1.15).

As a next example, let us consider the case when the Lagrangian does not depend explicitly on the independent variable x. In this case, we have the identity

$$\frac{d}{dx}\left(z_x \frac{\partial F}{\partial z_x} - F\right) = z_x \frac{d}{dx}\frac{\partial F}{\partial z_x} - \frac{\partial F}{\partial x} - \frac{\partial F}{\partial z}z_x$$

$$= -z_x\left(\frac{\partial F}{\partial z} - \frac{d}{dx}\frac{\partial F}{\partial z_x}\right) - \frac{\partial F}{\partial x} = 0, \qquad (1.102)$$

and therefore the expression in the first set of brackets above is a constant, i.e.,

$$z_x \frac{\partial F}{\partial z_x} - F = C. \qquad (1.103)$$

Taking this fact into account, one can conclude that the task of obtaining an extremal trajectory reduces to finding the solutions to the first order differential equation (1.103).

Example 1.8 As an illustration, let us consider the problem of finding the axially symmetric surface which has a minimal area enclosed by two co-axial rings. We choose the symmetry axis to coincide with the coordinate axis OX. The surface element shown in Fig. 1.16 has a length ds. The surface area confined between the two perpendicular planes passing through the points x and $x + dx$ of the symmetry axis is

$$d\mathcal{A} = 2\pi z(x)ds. \qquad (1.104)$$

If we express ds via the Cartesian coordinates, we can write

$$ds = (1 + z_x^2)^{\frac{1}{2}}dx,$$

and, respectively, for the infinitesimal surface area $d\mathcal{A}$, we have

Fig. 1.16 Axially symmetric surface connecting two co-axial circles which are perpendicular to the symmetry axis

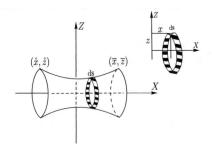

$$dA = 2\pi z(x)(1 + z_x^2)^{\frac{1}{2}} dx. \tag{1.105}$$

If the circles at the boundaries are located at the points $\mathring{x}$ and $\bar{x}$, then the full area is

$$dA = 2\pi \int_{\mathring{x}}^{\bar{x}} z(x)(1 + z_x^2)^{\frac{1}{2}} dx \tag{1.106}$$

and the integrand obviously does not depend explicitly on the independent variable x. Respectively, the Euler-Lagrange equation (1.103) has the form

$$z(1 + z_x^2)^{\frac{1}{2}} - z_x \frac{\partial}{\partial z_x}(z(1 + z_x^2)^{\frac{1}{2}}) = \text{const} \tag{1.107}$$

and, after some transformations, reduces to the equation

$$z(1 + z_x^2)^{\frac{1}{2}} - z z_x^2(1 + z_x^2)^{-\frac{1}{2}} = c, \tag{1.108}$$

in which c is a constant. By multiplying both sides of the above equation by $(1 + z_x^2)^{\frac{1}{2}}$, we obtain

$$z(1 + z_x^2)^{-\frac{1}{2}} = c, \tag{1.109}$$

and, respectively,

$$\frac{dx}{dz} = \frac{c}{\sqrt{z^2 - c^2}}. \tag{1.110}$$

Through a new integration, we end up with the formula

$$x = c \operatorname{arccosh} \frac{z}{c} + a, \tag{1.111}$$

in which a is the integration constant. Solving (1.111) for z ultimately gives us

$$z(x) = c \cosh \frac{x - a}{c}, \tag{1.112}$$

and one can easily realize that we are dealing with the profile curve generating the catenoid (see Fig. 1.16). As before, a further analysis is needed to prove that this is the real minimum, but we will omit this here. We shall mention, however, that this topic is discussed in greater detail in the book by Forsyth (1927).

Example 1.9 (Delaunay surfaces) Let us consider the rotational surface obtained by revolving the graph of the function $z(x)$ in the plane XOZ about the OX axis and look for the surface which encloses a fixed volume while having a minimal surface area. The volume V and the surface area $\mathcal{A}$ are given, respectively, by the formulas

$$V = \pi \int z^2(x)\mathrm{d}x, \qquad A = 2\pi \int z(x)\sqrt{1 + z_x^2(x)}\,\mathrm{d}x.$$

In order to find the extremal surface, we consider the functional

$$J = A - \lambda V = \int \left(2z(x)\sqrt{1 + z_x^2} - \lambda z^2(x) \right) \mathrm{d}x.$$

As the integrand F does not depend explicitly on the independent variable x, we can again use the integral (1.103) instead of the Euler-Lagrange equation, and this allows us to write the equation

$$2z\sqrt{1 + z_x^2} - \lambda z^2(x) - \frac{2zz_x(x)}{\sqrt{1 + z_x^2}} = c,$$

in which c is the integration constant. We can transform the above equation into the form

$$z^2(x) \pm \frac{2az(x)}{\sqrt{1 + z_x^2}} = \pm b^2,$$

in which $a = -1/\lambda$, $b = -\frac{c}{\lambda}$. Actually, this equation coincides with the one describing the generating curves of the Delaunay surfaces (see Eells 1987 and Oprea 2000).

Let us consider the case in which the integrand $F(x, z, z_x)$ is a total derivative with respect to x of the function $f(x, z)$, i.e.,

$$F(x, z, z_x) = \frac{\mathrm{d}f(x, z)}{\mathrm{d}x} = \frac{\partial f}{\partial x} + \frac{\partial f}{\partial z}z_x.$$

The Euler-Lagrange equation in this case is

$$\frac{\partial^2 f}{\partial z\partial x} + \frac{\partial^2 f}{\partial z^2}z_x - \frac{\mathrm{d}}{\mathrm{d}x}\frac{\partial f}{\partial z} = 0. \qquad (1.113)$$

The calculation of the total derivative with respect to x in the equation above confirms that it is identically satisfied as the condition

$$\frac{\partial^2 f}{\partial z\partial x} = \frac{\partial^2 f}{\partial x\partial z}$$

is always fulfilled.

This result also raises the question: which is the most general case in which the Euler-Lagrange equation is identically satisfied?

To answer the above question, let us write (1.94) in its expanded form

$$\frac{\partial f}{\partial z} - \frac{\partial^2 f}{\partial x \partial z_x} - \frac{\partial^2 f}{\partial z \partial z_x} z_x - \frac{\partial^2 f}{\partial z_x^2} z_{xx} = 0. \tag{1.114}$$

Since, in the first three terms on the left-hand side, the highest derivative with respect to z is of the first order, it will be identically satisfied provided that the coefficient in front of z_{xx} vanishes identically too. The last condition is equivalent to the statement that F is a linear function of z_x, i.e.,

$$F = p(x, z) + q(x, z)z_x. \tag{1.115}$$

The Euler-Lagrange equation (1.114) generated by this Lagrangian is

$$\frac{\partial p}{\partial z} + \frac{\partial q}{\partial z} z_x - \frac{dq}{dx} = \frac{\partial p}{\partial z} - \frac{\partial q}{\partial x} = 0, \tag{1.116}$$

and it is satisfied for all allowed values of the variables x and z. However, the last equality exactly expresses the condition that F of the type (1.115) is a total derivative $\frac{df}{dx}$ of some function f. So, we have actually proved that the necessary and sufficient condition for the Euler-Lagrange equation to be satisfied identically is that the Lagrangian should be chosen to be a total derivative $\frac{df}{dx}$ of some function $f(x, z)$. Implicit in this result is the fact that adding a total derivative to the Lagrangian does not change the Euler-Lagrange equation. This property is a direct consequence of the linearity of the Euler-Lagrange equation with respect to the integrand F.

1.4.3 Euler-Lagrange Equation for Two Independent Variables

If we try to apply the variational method to the problem of finding the extremal values of the area enclosed by a simple closed curve in the plane, we face the necessity of generalizing the Euler-Lagrange equation to the case of two independent variables. In such cases, we have to deal with an at least two times differentiable function F of the independent variables x and z, the function $w(x, z)$ and its first partial derivatives w_x and w_z. The problem is finding extremal values of the functional J associated with $F(x, z, w, w_x, w_z)$, i.e.,

$$J = \iint F(x, z, w, w_x, w_z)dxdz. \tag{1.117}$$

Generalizing the procedure in the one-dimensional case, we shall include the varied function in a one parameter family $\tilde{w}(x, z, \varepsilon)$ which attains its extremal value for $\varepsilon = 0$. We will define the function $\tilde{w}(x, z, \varepsilon)$, as before, as an infinitesimal deformation, i.e.,

$$\tilde{w}(x, z, \varepsilon) = \tilde{w}(x, z, 0) + \varepsilon\sigma(x, z) = \omega(x, z) + \varepsilon\sigma(x, z), \tag{1.118}$$

where $\sigma(x, z)$ is an arbitrary differentiable function of x and z, which vanishes on the boundary Γ of the plane domain $\mathcal{D}$. From (1.118), we directly obtain

$$\tilde{w}_x(x, z, \varepsilon) = \tilde{w}_x(x, z, 0) + \varepsilon\sigma_x(x, z)$$

and

$$\tilde{w}_z(x, z, \varepsilon) = \tilde{w}_z(x, z, 0) + \varepsilon\sigma_z(x, z).$$

Then, by differentiating the functional

$$J(\varepsilon) = \iint_{\mathcal{D}} F(x, z, \tilde{w}, \tilde{w}_x, \tilde{w}_z)\mathrm{d}x\mathrm{d}z \tag{1.119}$$

with respect to the infinitesimal parameter ε, we obtain

$$\frac{\partial J(\varepsilon)}{\partial \varepsilon} = \iint_{\mathcal{D}} \left(\frac{\partial F}{\partial w}\sigma + \frac{\partial F}{\partial w_x}\sigma_x + \frac{\partial F}{\partial w_z}\sigma_z \right) \mathrm{d}x\mathrm{d}z.$$

Now, integrating the last two terms under the integral by parts results in

$$\frac{\partial J(\varepsilon)}{\partial \varepsilon} = \iint_{\mathcal{D}} \left(\frac{\partial F}{\partial \tilde{w}} - \frac{\partial}{\partial x}\frac{\partial F}{\partial w_x} - \frac{\partial}{\partial z}\frac{\partial F}{\partial w_z} \right) \sigma(x, z)\mathrm{d}x\mathrm{d}z. \tag{1.120}$$

The expected extremal values of J are those for which

$$\frac{\partial J(\varepsilon)}{\partial \varepsilon}\Big|_{\varepsilon=0} = 0, \tag{1.121}$$

and in this way, from (1.120), we end up with the condition

$$\iint_{\mathcal{D}} \left(\frac{\partial F}{\partial \tilde{w}} - \frac{\partial}{\partial x}\frac{\partial F}{\partial \tilde{w}_x} - \frac{\partial}{\partial z}\frac{\partial F}{\partial \tilde{w}_z} \right) \sigma(x, z)\mathrm{d}x\mathrm{d}z = 0. \tag{1.122}$$

As $\sigma(x, z)$ is an arbitrary function, the last equation is satisfied if and only if

$$\frac{\partial F}{\partial \tilde{w}} - \frac{\partial}{\partial x}\frac{\partial F}{\partial \tilde{w}_x} - \frac{\partial}{\partial z}\frac{\partial F}{\partial \tilde{w}_z} = 0, \tag{1.123}$$

and this is desired for generalization of the Euler-Lagrange equation for the case of two independent variables.

Example 1.10 Let us consider a simple closed space curve $\mathcal{C}$ whose projection Γ on the plane $y \equiv 0$ is a simple closed curve and the surface $\mathcal{S}$ bounded by $\mathcal{C}$, which is of the form

$$y = w(x, z), \qquad (x, z) \in \mathcal{D}, \qquad \partial\mathcal{D} = \Gamma, \tag{1.124}$$

where $w(x, z)$ is the function which defines Γ (Monge parameterization). The area $\mathcal{A}$ of the surface $\mathcal{S}$ is given by the formula

$$\mathcal{A} = \iint_{\mathcal{D}} (1 + w_x^2 + w_z^2)^{1/2} dx dz. \tag{1.125}$$

According to (1.123), the extremum of $\mathcal{A}$ can be determined from solutions to the equation

$$\frac{\partial}{\partial x}\frac{\partial F}{\partial w_x} + \frac{\partial}{\partial z}\frac{\partial F}{\partial w_z} = 0, \tag{1.126}$$

where

$$F := (1 + w_x^2 + w_z^2)^{1/2}. \tag{1.127}$$

Let us recall the standard notation in the case of a Monge type parameterization namely,

$$p = w_x, \quad q = w_z, \quad \rho = 1 + p^2 + q^2$$
$$r = w_{xx}, \quad s = w_{xz}, \quad t = w_{zz}.$$

Using the notation above, we can write (1.126) in the form

$$\frac{\partial}{\partial x}\frac{p}{\rho} + \frac{\partial}{\partial z}\frac{q}{\rho} = 0, \tag{1.128}$$

and after a few differentiations, one ends up with

$$(1 + q^2)r - 2pqs + (1 + p^2)t = 0. \tag{1.129}$$

It is easy to recognize that the left side of (1.129) coincides (up to a nonzero multiplier) with the mean curvature H of the surface $\mathcal{S}$ in the Monge representation, and therefore (1.129) is the equation describing minimal surfaces in $\mathbb{R}^3$.

An alternative method for approaching the same problem is to use the parametric form, i.e., let J depend on two differentiable functions $x(t)$ and $z(t)$ of the independent variable t and their derivatives $\dot{x}(t)$ and $\dot{z}(t)$, i.e.,

$$J = \int F(t, x(t), z(t), \dot{x}(t), \dot{z}(t)) dt. \tag{1.130}$$

The variation of the last integral is now given by two independent variations of the functions $x(t) = \tilde{x}(t, 0)$ and $z(t) = \tilde{z}(t, 0)$

$$\tilde{x}(t, \varepsilon) = x(t) + \varepsilon\tau(t), \qquad \tau(\overset{\circ}{t}) = \tau(\bar{t}) = 0$$
$$\tilde{z}(t, \varepsilon) = z(t) + \varepsilon\sigma(t), \qquad \sigma(\overset{\circ}{t}) = \sigma(\bar{t}) = 0 \qquad (1.131)$$

and their derivatives

$$\dot{\tilde{x}}(t, \varepsilon) = \dot{x}(t) + \varepsilon\dot{\tau}(t)$$
$$\dot{\tilde{z}}(t, \varepsilon) = \dot{z}(t) + \varepsilon\dot{\sigma}(t). \qquad (1.132)$$

Following the same procedure as in the case of one variable, we have

$$\frac{\mathrm{d}J}{\mathrm{d}\varepsilon} = \int_{\overset{\circ}{t}}^{\bar{t}} \left(\frac{\partial F}{\partial\tilde{x}}\frac{\partial\tilde{x}}{\partial\varepsilon} + \frac{\partial F}{\partial\tilde{z}}\frac{\partial\tilde{z}}{\partial\varepsilon} + \frac{\partial F}{\partial\dot{\tilde{x}}}\frac{\partial\dot{\tilde{x}}}{\partial\varepsilon} + \frac{\partial F}{\partial\dot{\tilde{z}}}\frac{\partial\dot{\tilde{z}}}{\partial\varepsilon} \right) \mathrm{d}t$$

$$= \int_{\overset{\circ}{t}}^{\bar{t}} \left(\frac{\partial F}{\partial\tilde{x}}\tau + \frac{\partial F}{\partial\tilde{z}}\sigma + \frac{\partial F}{\partial\dot{\tilde{x}}}\dot{\tau} + \frac{\partial F}{\partial\dot{\tilde{z}}}\dot{\sigma} \right) \mathrm{d}t \qquad (1.133)$$

$$= \int_{\overset{\circ}{t}}^{\bar{t}} \left(\left(\frac{\partial F}{\partial\tilde{x}} - \frac{\mathrm{d}}{\mathrm{d}t}\frac{\partial F}{\partial\dot{\tilde{x}}} \right)\tau + \left(\frac{\partial F}{\partial\tilde{z}} - \frac{\mathrm{d}}{\mathrm{d}t}\frac{\partial F}{\partial\dot{\tilde{z}}} \right)\sigma \right) \mathrm{d}t.$$

At the point $\varepsilon = 0$, which corresponds to the extreme value of J, we have

$$\frac{\mathrm{d}J(\varepsilon)}{\mathrm{d}\varepsilon}\Big|_{\varepsilon=0} = \int_{\overset{\circ}{t}}^{\bar{t}} \left(\frac{\partial F}{\partial x} - \frac{\mathrm{d}}{\mathrm{d}t}\frac{\partial F}{\partial\dot{x}} \right)\tau\,\mathrm{d}t + \int_{\overset{\circ}{t}}^{\bar{t}} \left(\frac{\partial F}{\partial z} - \frac{\mathrm{d}}{\mathrm{d}t}\frac{\partial F}{\partial\dot{z}} \right)\sigma\,\mathrm{d}t = 0. \qquad (1.134)$$

The last equation is fulfilled for all values of τ and σ satisfying the boundary conditions, and as we can independently substitute $\tau = 0$ or $\sigma = 0$, this allows us to conclude that this time, we have two Euler-Lagrange equations

$$\frac{\partial F}{\partial x} - \frac{\mathrm{d}}{\mathrm{d}t}\frac{\partial F}{\partial\dot{x}} = 0, \qquad \frac{\partial F}{\partial z} - \frac{\mathrm{d}}{\mathrm{d}t}\frac{\partial F}{\partial\dot{z}} = 0. \qquad (1.135)$$

The above equations have been used in the analysis of the mylar balloon presented in the work of Hadzhilazova and Mladenov (2008).

Additionally, we would like to mention the following:

(1) The Eqs. (1.135) can be immediately generalized for an arbitrary number of independent functions $x(t), \ldots, z(t)$.
(2) As before, in the case in which the Lagrangian does not depend explicitly on the independent variable, i.e., $F = F(x(t), \ldots, z(t), \dot{x}(t), \ldots, \dot{z}(t))$, we have the first integral

$$\frac{\partial F}{\partial \dot{x}}\dot{x} + \dots + \frac{\partial F}{\partial \dot{z}}\dot{z} - F = C. \tag{1.136}$$

To prove this statement, it is enough to consider the identity

$$\frac{\mathrm{d}}{\mathrm{d}t}\left(\frac{\partial F}{\partial \dot{x}}\dot{x} + \dots + \frac{\partial F}{\partial \dot{z}}\dot{z} - F\right) = \left(\frac{\mathrm{d}}{\mathrm{d}t}\frac{\partial F}{\partial \dot{x}} - \frac{\partial F}{\partial x}\right)\dot{x} + \dots$$

$$+ \left(\frac{\mathrm{d}}{\mathrm{d}t}\frac{\partial F}{\partial \dot{z}} - \frac{\partial F}{\partial z}\right)\dot{z} - \frac{\partial F}{\partial t} = 0 \tag{1.137}$$

and to realize that the expressions in the brackets coincide with the Euler-Lagrange equations, while the last term on the right-hand side is zero by the initial assumption.

1.4.4   Euler-Lagrange Equation for Lagrangians Containing Second Order Derivatives

The case in which the integrand F contains second order derivatives with respect to the independent variable x, i.e., $F = F(x, \dot{z}(x), \ddot{z}(x))$, where $\dot{z} = \dfrac{\mathrm{d}z}{\mathrm{d}x}$, $\ddot{z} = \dfrac{\mathrm{d}^2 z}{\mathrm{d}x^2}$, can be treated in an analogous way to that presented in Sect. 1.4.1. It is easy to prove that the relevant Euler-Lagrange equation is

$$\frac{\mathrm{d}^2}{\mathrm{d}x^2}\frac{\partial F}{\partial \ddot{z}} - \frac{\mathrm{d}}{\mathrm{d}x}\frac{\partial F}{\partial \dot{z}} + \frac{\partial F}{\partial z} = 0. \tag{1.138}$$

This can be generalized for higher order derivatives (beyond the second, as is the case here) and for more than one dependent variable $z_1(x), z_2(x), \dots, z_n(x)$.

Correspondingly, in the parametric form, we will have

$$J = \int F(t, x(t), z(t), \dot{x}(t), \dot{z}(t), \ddot{x}(t), \ddot{z}(t))\mathrm{d}t,$$

and the relevant Euler-Lagrange equations are

$$\frac{\mathrm{d}^2}{\mathrm{d}t^2}\frac{\partial F}{\partial \ddot{x}} - \frac{\mathrm{d}}{\mathrm{d}t}\frac{\partial F}{\partial \dot{x}} + \frac{\partial F}{\partial x} = 0$$

$$\frac{\mathrm{d}^2}{\mathrm{d}t^2}\frac{\partial F}{\partial \ddot{z}} - \frac{\mathrm{d}}{\mathrm{d}t}\frac{\partial F}{\partial \dot{z}} + \frac{\partial F}{\partial z} = 0. \tag{1.139}$$

References

E. Cesáro, *Lezioni di Geometria Intrinseca* (Presso L'Autore-Editore, Naples, 1896)

J. Eells, The surfaces of Delaunay. Math. Intell. **9**, 53–57 (1987)

L. Eisenhart, *A Treatise on the Differential Geometry of Curves and Surfaces* (Ginn and Co., Boston, 1909)

A. Forsyth, *Calculus of Variations* (Cambridge University Press, Cambridge, 1927)

M. Hadzhilazova, I. Mladenov, Once more the mylar balloon. CRAS (Sofia) **61**, 847–856 (2008)

J. McCleary, *Geometry From a Differentiable Viewpoint* (Cambridge University Press, Cambridge, 1994)

J. Oprea, *The Mathematics of Soap Films: Explorations with Maple®* (AMS, Providence, 2000)

J. Oprea, *Differential Geometry and Its Applications*, 3rd edn. (Mathematical Association of America, Washington D. C, 2007)

D. Struik, *Lectures on Classical Differential Geometry* (Dover, New York, 1988)

W. Whewell, On the intrinsic equation of a curve, and its application. Camb. Phil. Soc. **8**, 659–671 (1849)

Chapter 2
Planar Curves Whose Curvature Depends Only on the Distance From a Fixed Point

Abstract Looking at the Frenet-Serret equations from the viewpoint of dynamical systems, one can prove that when the curvature of a plane curve is given as a function of the radius, the problem of reconstructing this curve is reducible to quadratures. Additionally, two different integration procedures are presented. These methods are illustrated first via the famous lemniscate of Bernoulli, which is immediately related to the Euler elastica. Relying on the new formalism, the Sturm spirals and their generalizations, the Serret curves (which have a mechanical origin) and their generalizations are parametrized explicitly. The results on the Serret curves are original, as their description up to now has been purely abstract. Finally, the same technique is applied to the Cassinian ovals, and in this way, one concludes with their alternative parameterizations.

2.1 The Moving Frame Associated with a Plane Curve

The fundamental existence and uniqueness theorem in the theory of plane curves states that a curve is uniquely determined (up to Euclidean motion) by its curvature given as a function of its arc-length (see Berger and Gostiaux (1988), p. 296 or Oprea (2007), p. 37). The simplicity of the situation, however, is elusive, as in many cases, it is impossible to find the curve explicitly. Having that in mind, it is clear that if the curvature is given as a function of its position, the situation is even more complicated. Viewing the Frenet-Serret equations as a fictitious dynamical system, it was proven in Vassilev et al. (2009) (see also Djondjorov et al. (2009a)) that when the curvature is given simply as a function of the distance from the origin, the problem can always be reduced to quadratures. This last result is not entirely new, as Singer (1999) had already shown that in some cases, it is possible that such a curvature has an interpretation as a central potential in the plane, and therefore the trajectories can be found through the standard procedures in classical mechanics. However, the approach which we will follow here is entirely different from the group-theoretical approach of Vassilev et al. (2009) or the mechanical approach of Singer (1999) proposed in those papers. The method is illustrated by the most natural example in the class of curves whose curvatures are functions only of the distance from the

© Springer International Publishing AG 2017
I.M. Mladenov and M. Hadzhilazova, *The Many Faces of Elastica*,
Forum for Interdisciplinary Mathematics 3, DOI 10.1007/978-3-319-61244-7_2

origin. Here, we consider in some detail the cases in which the function in question is either proportional or inversely proportional to the distance from the origin. Let us start with the first case, namely,

$$\kappa = \sigma r, \quad r = |\mathbf{x}| = \sqrt{x^2 + z^2}, \tag{2.1}$$

where x, z are the Cartesian coordinates in the plane XOZ, which have to be considered as functions of the arc-length parameter s, and σ is assumed to be a positive real constant.

If $\theta(s)$ denotes the slope of the tangent to the curve with respect to the OX axis, one has the following geometrical relations:

$$\frac{d\theta(s)}{ds} = \kappa(s), \quad \frac{dx}{ds} = \cos\theta(s), \quad \frac{dz}{ds} = \sin\theta(s), \tag{2.2}$$

which can also be deduced from the Frenet-Serret equations (see also Fig. 1.1)

$$\frac{d\mathbf{x}(s)}{ds} = \mathbf{T}(s), \quad \frac{d\mathbf{T}}{ds} = \kappa\mathbf{N}, \quad \frac{d\mathbf{N}}{ds} = -\kappa\mathbf{T}, \tag{2.3}$$

where $\mathbf{T}$ and $\mathbf{N}$ are, respectively, the tangent and the normal vectors to the curve, and s is the natural parameter along it. Combining (2.1) and (2.2), we get

$$\frac{d\theta(s)}{ds} = \kappa(r), \tag{2.4}$$

which is still quite an unpromising equation. We will proceed (as suggested but not pursued in Singer (1999)) by going to the co-moving frame $(\mathbf{T}, \mathbf{N})$ associated with the curve

$$\mathbf{x} = \xi\mathbf{T} + \eta\mathbf{N} \tag{2.5}$$

accordingly, the Frenet-Serret equations (2.3) take the form

$$\frac{d\xi}{ds} = \dot{\xi} = \kappa\eta + 1, \quad \frac{d\eta}{ds} = \dot{\eta} = -\kappa\xi. \tag{2.6}$$

2.2   Integration

Multiplying the first equation in (2.6) by ξ, the second one by η, and summing up the so-obtained expressions, we find that

$$\xi = r\dot{r}, \tag{2.7}$$

in which the dot means a differentiation with respect to the arc-length parameter. Substituting this expression back into the second equation of (2.6) and integrating, we obtain

$$\eta = -\int \kappa(r)rdr + c, \tag{2.8}$$

where c is the integration constant. One should notice, however (cf. equation (2.5)), that the coordinates in the moving frame are not entirely independent, but rather obey the constraint

$$\xi^2 + \eta^2 = r^2, \tag{2.9}$$

which, in view of Eqs. (2.7) and (2.8), presents an ordinary differential equation for the radial coordinate r.

2.3   Bernoulli's Lemniscates

This curve is a special case (when $a \equiv c$) of the Cassinian ovals (see Mladenov (2000) and the end of this chapter), defined by the equation

$$(x^2 + z^2)^2 - 2a^2(z^2 - x^2) + a^4 - c^4 = 0, \tag{2.10}$$

and has a curvature (which can be found using formula (1.18)) that is linear in r. Inserting $\kappa = \sigma r$ into Eq. (2.8) produces

$$\eta = -\frac{\sigma r^3}{3} \tag{2.11}$$

(the integration constant is taken to be zero) and the scheme from the previous section leads to the equation

$$\frac{dr}{ds} = \sqrt{1 - \frac{\sigma^2 r^4}{9}}. \tag{2.12}$$

Its integration is immediate and gives us

$$r = \sqrt{\frac{3}{\sigma}} \, \mathrm{cn}(\sqrt{\frac{2\sigma}{3}}s, \frac{1}{\sqrt{2}}), \tag{2.13}$$

where $\mathrm{cn}(u, k)$ denotes one of the Jacobian elliptic functions in which the first slot is occupied by its argument and the second one by the so-called elliptic modulus (a real number between zero and one). More details about elliptic functions and integrals can be found in Jahnke et al. (1960) and Olver et al. (2010). Substituting this solution into Eqs. (2.7) and (2.8) has, as a result, the coordinates of the lemniscate in the moving frame

$$\xi = -\sqrt{\frac{6}{\sigma}} \operatorname{cn}(\sqrt{\frac{2\sigma}{3}}s, \frac{1}{\sqrt{2}}) \operatorname{dn}(\sqrt{\frac{2\sigma}{3}}s, \frac{1}{\sqrt{2}}) \operatorname{sn}(\sqrt{\frac{2\sigma}{3}}s, \frac{1}{\sqrt{2}})$$

$$(2.14)$$

$$\eta = -\sqrt{\frac{3}{\sigma}} \operatorname{cn}^3(\sqrt{\frac{2\sigma}{3}}s, \frac{1}{\sqrt{2}}).$$

With respect to the fixed one, these functions give a new curve which we will call the co-lemniscate. Written in terms of its components, Eq. (2.5) tells us that the lemniscate coordinates x, z are obtained from those of the co-lemniscate ξ, η via a plane rotation specified by the slope angle θ, i.e.,

$$x = \xi \cos \theta - \eta \sin \theta, \qquad z = \xi \sin \theta + \eta \cos \theta. \qquad (2.15)$$

Obviously, what remains to be done is to find θ, and this can be achieved via an integration of the first equation in (2.2). In this way, we obtain

$$\theta = 3 \arccos(\operatorname{dn}(\sqrt{\frac{2\sigma}{3}}s, \frac{1}{\sqrt{2}}))$$

$$(2.16)$$

$$= 3 \arcsin(\tilde{k} \operatorname{sn}(\sqrt{\frac{2\sigma}{3}}s, \frac{1}{\sqrt{2}})), \qquad \tilde{k} = \sqrt{1 - k^2}.$$

Now we have to take into account the trigonometric identities

$$\sin 3\varphi = 3 \sin \varphi - 4 \sin^3 \varphi, \qquad \cos 3\varphi = 4 \cos^3 \varphi - 3 \cos \varphi, \qquad (2.17)$$

which give us

$$\sin \theta = \frac{3}{\sqrt{2}} \operatorname{sn}(\sqrt{\frac{2\sigma}{3}}s, \frac{1}{\sqrt{2}}) - \sqrt{2} \operatorname{sn}^3(\sqrt{\frac{2\sigma}{3}}s, \frac{1}{\sqrt{2}})$$

$$(2.18)$$

$$\cos \theta = 4\operatorname{dn}^3(\sqrt{\frac{2\sigma}{3}}s, \frac{1}{\sqrt{2}}) - 3\operatorname{dn}(\sqrt{\frac{2\sigma}{3}}s, \frac{1}{\sqrt{2}})$$

and eventually provide the parameterization of the Bernoullian lemniscate. By making repeated use of the fundamental identities which the Jacobian elliptic functions $\operatorname{sn}(u, k)$, $\operatorname{cn}(u, k)$ and $\operatorname{dn}(u, k)$ satisfy, i.e.,

$$\operatorname{sn}^2(u, k) + \operatorname{cn}^2(u, k) = 1, \qquad \operatorname{dn}^2(u, k) + k^2 \operatorname{sn}^2(u, k) = 1, \qquad (2.19)$$

it is possible to simplify the expressions for x and z into the form

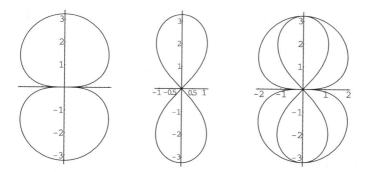

Fig. 2.1 The Bernoullian co-lemniscate (*left*), Bernoulli's lemniscate (*middle*) and both of them (*right*) drawn via formulas (2.14) and (2.20) with $\sigma = 3.5$

$$x = \sqrt{\frac{3}{2\sigma}}\mathrm{cn}(\sqrt{\frac{2\sigma}{3}}s, \frac{1}{\sqrt{2}})\mathrm{sn}(\sqrt{\frac{2\sigma}{3}}s, \frac{1}{\sqrt{2}})$$

$$z = -\sqrt{\frac{3}{\sigma}}\mathrm{cn}(\sqrt{\frac{2\sigma}{3}}s, \frac{1}{\sqrt{2}})\mathrm{dn}(\sqrt{\frac{2\sigma}{3}}s, \frac{1}{\sqrt{2}}).$$

(2.20)

The properties of the Jacobian functions also make obvious the relations

$$\eta^2 = \left(\frac{\sigma}{3}\right)^2 (\xi^2 + \eta^2)^3, \qquad z^2 - x^2 = \frac{\sigma}{3}\left(x^2 + z^2\right)^2,$$

(2.21)

the last one being simply the standard form of Bernoulli's lemniscate (cf. the equation (2.10)). In polar coordinates $\xi = r\cos\psi, \eta = r\sin\psi, x = r\cos\phi, z = r\sin\phi$, the above algebraic curves are of degree six and four, respectively, and take the form

$$\sin\psi = -\frac{\sigma}{3}r^2, \qquad \cos 2\phi = -\frac{\sigma}{3}r^2.$$

(2.22)

This remarkable similarity in their polar representations suggests calculation of the curvature of the co-lemniscate. The most convenient method in this situation seems to be the application of the formula

$$\kappa = \frac{|r^2 + 2\dot{r}^2 - r\ddot{r}|}{(r^2 + \dot{r}^2)^{3/2}},$$

(2.23)

in which this time the dots denote differentiations with respect to the polar angle. Through straightforward but tedious calculations, one concludes with the formula

$$\kappa = \frac{2\sigma r\left(\sigma^2 r^4 + 9\right)}{\sqrt{3}\left(\sigma^2 r^4 + 3\right)^{3/2}}.$$

(2.24)

Regardless of how similar this curve seems to be to the lemniscate in the polar coordinates, its curvature is quite different from that of the parent curve. Both curves are plotted for illustration in (Fig. 2.1).

Remark Because of the relation between the radial coordinate r and the curvature equation (2.1), we can rewrite Eq. (2.12) in the form

$$\dot{\kappa}^2 + \frac{\kappa^4}{9} = \sigma^2, \tag{2.25}$$

which can be recognized and is referred to further on as the *intrinsic equation* of the Bernoullian lemniscate.

2.4 Relationship Between the Lemniscate and the Elastica

Here, we will outline the relation of Bernoulli's lemniscate to another famous curve invented by Bernoulli, the so-called free (or rectangular) elastica (cf. Djondjorov et al. (2009)), which also appears as a profile curve of the Mylar balloon described in detail in Hadzhilazova and Mladenov (2008) and Mladenov and Oprea (2009).

For this purpose, let us differentiate Eq. (2.25), which gives us

$$\ddot{\kappa}_{\text{lemn}} + \frac{2}{9}\kappa_{\text{lemn}}^3 = 0 \tag{2.26}$$

and presents another form of the intrinsic equation of this interesting curve. By comparing it with the intrinsic equation of the free elastica

$$\ddot{\kappa}_{\text{elas}} + \frac{1}{2}\kappa_{\text{elas}}^3 = 0, \tag{2.27}$$

it is easy to conclude that they are related by the transformation

$$\kappa_{\text{lemn}} = \frac{3}{2}\kappa_{\text{elas}}, \tag{2.28}$$

which has been noticed recently by Matsutani (2010) as well. Actually, the mathematical reason is that the curvature of the Bernoulli elastica is again a linear function of the distance, but this time from the OX axis.

2.5 Spirals

In their article on the subject, Bourbaki have called them the most mysterious curves. Actually, we will not unveil their secrets here, but will simply prove some of their properties, starting with the fundamental geometrical characteristic that the spirals are just the curves for which the curvature is inversely proportional to the distance from a fixed point in the plane called a pole. In analytical form, this property is expressed via the formula

$$\kappa = \frac{\sigma}{|\mathbf{x}|} = \frac{\sigma}{r} = \frac{\sigma}{\sqrt{x^2 + z^2}}, \qquad \sigma > 0, \tag{2.29}$$

where x, z are the Cartesian coordinates in the plane XOZ, which have to be considered as functions of the arc-length parameter s, and σ is assumed to be a positive real constant. Depending on the numerical value of this constant, we will consider several types of spiral which we will describe in detail.

2.6 Sturmian Spirals

By their very definition (cf. Zwikker (1963)), these plane curves possess the property that at each point, their curvature radius $\mathcal{R}$ coincides with the distance r from the origin. Formulated in curvature terms, this means that their curvature κ is given by formula (2.1), in which $\sigma \equiv 1$. Applying the scheme from Sect. 2.2, one easily finds that

$$\eta = -r + c \tag{2.30}$$

and

$$\dot{r} = \frac{\sqrt{2cr - c^2}}{r}, \qquad c > 0. \tag{2.31}$$

It is convenient to perform the integration of the above equation by switching to a new independent variable t defined by the equation

$$\frac{ds}{dt} = r. \tag{2.32}$$

This leads to the following results:

$$r = \frac{c}{2}\left(t^2 + 1\right), \qquad \xi = ct, \qquad \eta = \frac{c}{2}\left(1 - t^2\right). \tag{2.33}$$

Integration of the first equation in (2.2) additionally leads to the new parameter t coinciding (up to a real constant) with the slope angle, i.e.,

$$\theta = t. \tag{2.34}$$

By rewriting Eq. (2.5) in its components, one also has the relations

$$x = \xi \cos t - \eta \sin t, \qquad z = \eta \cos t + \xi \sin t, \tag{2.35}$$

which, combined with the above findings, provides the sought after parameterization of the Sturmian spirals

$$x = c\left(t \cos t + \frac{t^2 - 1}{2} \sin t\right), \qquad z = c\left(\frac{1 - t^2}{2} \cos t + t \sin t\right). \tag{2.36}$$

Making use of the above formulas, one also easily finds the arc-length as a function of the parameter t, i.e.,

$$s = \frac{c}{2}\left(\frac{t^3}{3} + t\right). \tag{2.37}$$

By exchanging the numerical parameter c for 2ρ and taking into account the fundamental relation $\mathcal{R} \equiv r$ (for this curve), the above formula can be written into the following form, which is none other than the intrinsic equation (in Cesáro form) of the Sturm spiral:

$$s = \frac{(\mathcal{R} + 2\rho)}{3} \sqrt{\frac{\mathcal{R} - \rho}{\rho}}. \tag{2.38}$$

2.7 Generalized Sturm Spirals

Due to the restriction on the allowed values of σ, one can also consider the other two obvious possibilities, $\sigma > 1$ and $0 < \sigma < 1$, which have to be viewed as a generalization of the ordinary Sturmian spirals.

2.7.1 The Case When $\sigma > 1$

Here, we will just outline the main ingredients of the derivation, again following the scheme described in Sect. 2.2 starting with the equations

$$\eta = -\sigma r + c \quad \text{and} \quad \frac{dr}{ds} = \frac{\sqrt{(1 - \sigma^2)r^2 + 2c\sigma r - c^2}}{r}. \tag{2.39}$$

One easily concludes that the expression under the radical on the right-hand side is positive, provided that $c > 0$ and r belongs either to a finite or infinite interval, i.e.,

$$\frac{c}{\sigma + 1} \le r \le \frac{c}{\sigma - 1} \quad \text{and} \quad \sigma > 1 \quad \text{or} \quad r > \frac{c}{\sigma + 1} \quad \text{and} \quad \sigma < 1. \tag{2.40}$$

As the subsection title suggests, our immediate task is to consider the first of the possibilities presented above. Exchanging, as before, the arc-length parameter (cf. Eq. (2.32)) with t, leads to the formula

$$r = \frac{c}{\sigma^2 - 1}(\sigma + \sin\sqrt{\sigma^2 - 1}\,t), \qquad t \in [-\frac{\pi}{2\sqrt{\sigma^2 - 1}}, \frac{\pi}{2\sqrt{\sigma^2 - 1}}], \qquad (2.41)$$

by which we also find

$$\xi = \frac{dr}{dt} = \frac{c}{\sqrt{\sigma^2 - 1}}\cos\sqrt{\sigma^2 - 1}\,t, \qquad \theta = \sigma t. \qquad (2.42)$$

The combination of the above results with those from Eq. (2.41), the first equation in (2.39) and the general relations (2.35) gives us

$$x = c\left(\frac{\cos\sqrt{\sigma^2 - 1}\,t\,\cos\sigma t}{\sqrt{\sigma^2 - 1}} + \frac{(\sigma\sin\sqrt{\sigma^2 - 1}\,t + 1)\sin\sigma t}{\sigma^2 - 1}\right)$$

$$\qquad (2.43)$$

$$z = c\left(\frac{\cos\sqrt{\sigma^2 - 1}\,t\,\sin\sigma t}{\sqrt{\sigma^2 - 1}} - \frac{\left(\sigma\sin\sqrt{\sigma^2 - 1}\,t + 1\right)\cos\sigma t}{\sigma^2 - 1}\right).$$

The expressions for the arc-length and the intrinsic equation in this case are

$$s = \frac{c}{\sigma^2 - 1}(\sigma t - \frac{\cos\sqrt{\sigma^2 - 1}\,t}{\sqrt{\sigma^2 - 1}} + \frac{\sigma\pi}{2\sqrt{\sigma^2 - 1}}) \qquad (2.44)$$

and

$$s = \frac{c}{\sigma^2 - 1}\left(\frac{\sigma}{\sqrt{\sigma^2 - 1}}\arcsin(\frac{\sigma}{c}\left((\sigma^2 - 1)\mathcal{R} - c\right))\right.$$

$$\qquad (2.45)$$

$$\left. -\frac{1}{c}\sqrt{\sigma^2(1 - \sigma^2)\mathcal{R}^2 + 2c\sigma^2\mathcal{R} - c^2} + \frac{\sigma\pi}{2\sqrt{\sigma^2 - 1}}\right),$$

where, in the derivation of the last equation, we have used the defining relation for the spiral, which in this case states that $r = \sigma\mathcal{R}$.

A few remarks are in order here. First, while r takes its values in the interval (2.40), the variable t is running in the interval $[-\frac{\sigma\pi}{2\sqrt{\sigma^2-1}}, \frac{\sigma\pi}{2\sqrt{\sigma^2-1}}]$, and during this excursion, the tangent to the curve turns to the angle $\frac{\sigma\pi}{\sqrt{\sigma^2-1}}$. This angle is greater than 2π for $\sigma < \frac{2}{\sqrt{3}}$, equal to 2π for $\sigma = \frac{2}{\sqrt{3}}$ and less than 2π for $\sigma > \frac{2}{\sqrt{3}}$. All this is illustrated in Fig. 2.2.

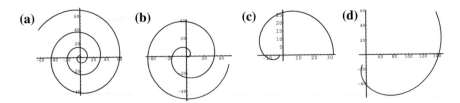

Fig. 2.2 a The standard Sturmian spiral generated by (2.36) and $c = 0.25$, and the generalized Sturmian spirals drawn via formulas in (2.43) with the following set of parameters: **b** $c = 1$, $\sigma = 1.02$, **c** $c = 5$, $\sigma = \frac{2}{\sqrt{3}}$, **d** $c = 100$, $\sigma = 5/3$

2.7.2 The Case When $0 < \sigma < 1$

The first steps in the scheme amount to

$$\eta = -\sigma r + c, \qquad \frac{dr}{ds} = \frac{\sqrt{(1 - \sigma^2)r^2 + 2c\sigma r - c^2}}{r}, \tag{2.46}$$

but one should keep in mind that now $\sigma < 1$ and $r > \frac{c}{\sigma+1}$. It also turns out to be more convenient to perform the integration of the equation on the right-hand side in (2.46) by introducing the parameter τ via the equation

$$\frac{ds}{d\tau} = r^2, \tag{2.47}$$

which produces

$$r = \frac{c}{\sigma - \sin c\tau}, \qquad \tau \in [-\frac{\pi}{2c}, \frac{\arcsin \sigma}{c}] \tag{2.48}$$

and

$$\xi = -\frac{c \cos c\tau}{\sigma - \sin c\tau}, \qquad \eta = -\frac{c \sin c\tau}{\sigma - \sin c\tau}$$

$$\tag{2.49}$$

$$\theta(\tau) = \frac{\sigma}{\sqrt{1 - \sigma^2}} \ln \frac{\sigma \tan \frac{c\tau}{2} - \sqrt{1 - \sigma^2} - 1}{\sigma \tan \frac{c\tau}{2} + \sqrt{1 - \sigma^2} - 1}.$$

Furthermore, via Eqs. (2.35) and (2.49), we obtain

$$x = \frac{c}{\sigma - \sin c\tau} \cos \left(c\tau - \frac{\sigma}{\sqrt{1-\sigma^2}} \ln \frac{\sigma \tan \frac{c\tau}{2} - \sqrt{1-\sigma^2} - 1}{\sigma \tan \frac{c\tau}{2} + \sqrt{1-\sigma^2} - 1} \right)$$

$$(2.50)$$

$$z = \frac{c}{\sigma - \sin c\tau} \sin \left(c\tau - \frac{\sigma}{\sqrt{1-\sigma^2}} \ln \frac{\sigma \tan \frac{c\tau}{2} - \sqrt{1-\sigma^2} - 1}{\sigma \tan \frac{c\tau}{2} + \sqrt{1-\sigma^2} - 1} \right),$$

and finally

$$s = \frac{c\sigma}{\left(1-\sigma^2\right)^{3/2}} \ln \left(\frac{\sigma + (1+\sqrt{1-\sigma^2}) \tan \left(\frac{1}{2} \arcsin \left(\frac{c}{r} - \sigma \right) \right)}{1 + \sqrt{1-\sigma^2} + \sigma \tan \left(\frac{1}{2} \arcsin \left(\frac{c}{r} - \sigma \right) \right)} \right)$$

$$(2.51)$$

$$+ \frac{\sqrt{(1-\sigma^2)r^2 + 2c\sigma r - c^2}}{1-\sigma^2}.$$

As before, one can easily obtain the intrinsic equation of the curve from the last expression by replacing r with $\sigma\mathcal{R}$ (Fig. 2.3).

2.7.3 The Sub-case When $0 < \sigma < 1$ and $c = 0$

Just for the sake of completeness we will consider the situation when the integration constant c, which appears in the previous subsection, is zero. Obviously, the equations in (2.46) simplify to

$$\eta = -\sigma r, \qquad \frac{dr}{ds} = \sqrt{1-\sigma^2}.$$

$$(2.52)$$

The integration of the second one is immediate and gives us

$$r = \sqrt{1-\sigma^2}\, s + a,$$

$$(2.53)$$

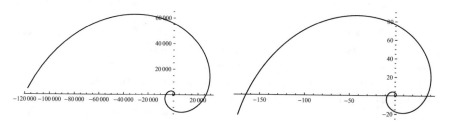

Fig. 2.3 Sturmian spirals generated by formula (2.50) and the constants $\sigma = 0.9$, $c = 1$ (*left*) and formula (2.56) with $\sigma = 0.9$ and $a = 1$ (*right*)

where a denotes a new integration constant that is necessarily positive. Following the scheme, one ends with the results

$$\xi = \left(1 - \sigma^2\right)s + a\sqrt{1 - \sigma^2}, \qquad \eta = -\sigma\left(\sqrt{1 - \sigma^2}\,s + a\right) \qquad (2.54)$$

$$\theta(s) = \frac{\sigma}{\sqrt{1 - \sigma^2}}\ln(\sqrt{1 - \sigma^2}\,s + a), \qquad (2.55)$$

which allow us to write down the explicit parameterization of the corresponding spiral

$$x = ((1 - \sigma^2)s + a\sqrt{1 - \sigma^2})\cos\theta(s) + \sigma(\sqrt{1 - \sigma^2}\,s + a)\sin\theta(s)$$
$$(2.56)$$
$$z = ((1 - \sigma^2)s + a\sqrt{1 - \sigma^2})\sin\theta(s) - \sigma(\sqrt{1 - \sigma^2}\,s + a)\cos\theta(s).$$

Remark It was a challenging task for the present authors (see Mladenov et al. (2011a)) to provide this detailed description from the first principles for the family of Sturm spirals that have such nice geometrical characteristics and to supply their explicit parameterizations. It was kind of a surprise to learn that they do not coincide with any of the famous spirals bearing the names of Archimedes, Cotes, Euler (Cornu), Fermat, Galileo, Nielsens, Poinsot, etc., that are traditionally covered in the textbooks on classical differential geometry (Berger and Gostiaux (1988), Doss-Bachelet et al. (2000), Gray (1998), Oprea (2007) and Rovenski (2000)). Only in the book by Zwikker (1963) was a short note found about the Norwich or Sturm spirals discussed in the previous section. A more thorough search, however, clarifies that the equiangular spiral

$$x = m\,e^{\phi\cot\alpha}\cos\phi, \qquad z = m\,e^{\phi\cot\alpha}\sin\phi, \qquad m, \alpha \in \mathbb{R}^+, \quad \phi \in \mathbb{R}, \qquad (2.57)$$

discovered by Descartes and sometimes also called a logarithmic spiral, possesses a curvature given by the formula $\kappa = \sin\alpha/r$, and therefore belongs to the class studied in the last two sections. The name of the curve comes from the fact that it cuts the radius vectors from the origin at a constant angle α. This also seems to be the reason behind the fact that insects approach candles along this curve, thinking perhaps that they are flying along a straight line at a constant angle to the rays of the light. Among other interesting properties of this curve, we mention that successive generation of its evolutes, pedal curves or inverses are still equiangular spirals (for more details, see Yates (1959)).

2.8   Serret Curves

A long time ago, Serret (1845) described a family of plane algebraic curves in response to a question raised by Legendre. The problem was to find algebraic curves other than the lemniscate, such that their arc lengths are expressed by elliptic integrals of the first kind. Serret claimed that he had found all such rational curves. Moreover, he provided a mechanical procedure for their construction (Serret (1845a)), which will be described below.

Before that, we will mention that the original Serret curves were indexed by natural numbers, but Liouville (1845) recognized immediately that rational numbers are well-suited enough, as they also lead to algebraic curves. This has been further elucidated in Krohs (1891) dissertation. Here, we extend the definition of Serret's curves from a discrete to a continuous two-parameter family and present their explicit parameterizations. Serret's curves were introduced as a trace of the end point M of the segment OM in the plane XOZ shown in Fig. 2.4.

The lengths of the hinged rods OP and PM are specified by a natural number $n \in \mathbb{N}$ via the formulas $\sqrt{n}$ and $\sqrt{n+1}$, while the point O is fixed at the origin of the Cartesian coordinate system XOZ.

During its movement, the point M describes the curve S_n according to the rule

$$\cos \omega = \cos(n\alpha - (n+1)\beta), \tag{2.58}$$

where the angles α, β and ω are shown in Fig. 2.4. Their analytical treatment is based on the application of the cosine theorem to the triangle OMP, which gives us

$$\cos \alpha = \frac{r^2 - 1}{2\sqrt{n}r}, \qquad \cos \beta = \frac{r^2 + 1}{2\sqrt{n+1}r}. \tag{2.59}$$

Following Liouville's observation, we obtain an algebraic curve even after replacing the index n with the rational number $\nu = p/q$, where $p, q \in \mathbb{N}$. This can be seen to be true by expressing the right-hand side of (2.58) as a polynomial in $\sin \alpha$, $\cos \alpha$, $\sin \beta$ and $\cos \beta$, and then by making use of the geometrical relations (2.59) to obtain the algebraic relations between x and r, and z and r in the form of polynomial equations, i.e.,

Fig. 2.4   Serret construction

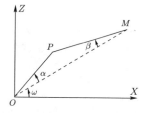

$$P(x, r) = 0, \qquad Q(z, r) = 0. \tag{2.60}$$

Eliminating r between them, one concludes with some polynomial relation

$$F(x, z) = 0, \tag{2.61}$$

and this proves that the curve S_ν is algebraic.

We should note that Lipkovski (1996) was recently able to prove that all Serret curves S_n are rational ones, i.e., they admit rational parameterizations.

Going back to the original Serret (1845a) writings, one can find a formula for the curvature of S_n in the form

$$\kappa(r) = \frac{3r}{2\sqrt{n(n+1)}} - \frac{2n+1}{2\sqrt{n(n+1)}r}, \tag{2.62}$$

which depends solely on the radial coordinate r. This will be used in the next section to generate them by following the original construction of Mladenov et al. (2010, 2012).

2.8.1 Generalized Serret Curves

The expression for the curvature of Serret's curves (2.62) immediately suggests a generalization of the form

$$\kappa(r) = 3\lambda r - \frac{\sigma}{r}, \qquad \lambda > 0, \qquad \sigma > 1. \tag{2.63}$$

Substitution of (2.63) into (2.8) produces

$$\eta = -\lambda r^3 + \sigma r, \tag{2.64}$$

but one has to notice that the integration constant in (2.8) is taken to be zero. Under these circumstances, the differential equation

$$\dot{r}^2 = \frac{1}{r^2} \left(r^2 - \eta^2 \right), \tag{2.65}$$

which follows from (2.9),reduces to the equation

$$\frac{dr}{\sqrt{(a^2 - r^2)(r^2 - c^2)}} = \lambda ds, \tag{2.66}$$

in which the real parameters a and c are given by the formulas

$$a = \sqrt{\frac{\sigma + 1}{\lambda}}, \qquad c = \sqrt{\frac{\sigma - 1}{\lambda}}. \tag{2.67}$$

The integration of (2.66) can be performed in terms of the Jacobian elliptic function $\mathrm{dn}(\cdot, \cdot)$, namely,

$$r(s) = a\,\mathrm{dn}(a\lambda s, k), \qquad k = \sqrt{\frac{2}{\sigma + 1}}, \tag{2.68}$$

The next step in the scheme amounts to evaluation of the integral

$$\theta(s) = \int \kappa(r(s))\mathrm{d}s, \tag{2.69}$$

and this gives us

$$\theta(s) = 3\,\mathrm{am}(\sqrt{\lambda(\sigma + 1)}\,s, k) - \frac{\sigma}{\sqrt{\sigma^2 - 1}}\arccos\frac{\mathrm{cn}(\sqrt{\lambda(\sigma + 1)}\,s, k)}{\mathrm{dn}(\sqrt{\lambda(\sigma + 1)}\,s, k)}, \tag{2.70}$$

where $\mathrm{am}(t, k)$ is the Jacobian amplitude function and $\mathrm{cn}(t, k) = \cos \mathrm{am}(t, k)$.

Having at our disposal (2.7), (2.64), (2.68) and (2.70) can enter into (2.15), giving us the parameterization of the generalized Serret curves. Obviously, the parameterization of the classical Serret curves can be obtained by taking

$$\lambda = \frac{1}{2\sqrt{n(n + 1)}}, \qquad \sigma = \frac{2n + 1}{2\sqrt{n(n + 1)}}, \qquad n \in \mathbb{N}, \tag{2.71}$$

and in this case, the slope angle turns out to be

$$\theta_n(s) = 3\,\mathrm{am}(\mu_n s, k_n) - (2n + 1)\arccos\frac{\mathrm{cn}(\mu_n s, k_n)}{\mathrm{dn}(\mu_n s, k_n)}, \tag{2.72}$$

where

$$\mu_n = \frac{1}{2}\sqrt{\frac{2\sqrt{n(n + 1)} + 2n + 1}{n(n + 1)}}, \qquad k_n = 2\sqrt{\frac{\sqrt{n(n + 1)}}{2\sqrt{n(n + 1)} + 2n + 1}}. \tag{2.73}$$

Several plots of both classical and generalized Serret curves are presented in Fig. 2.5 and Fig. 2.6.

Remark From the viewpoint of curve engineering, the curvature in (2.63) is a superposition of Bernoulli's lemniscate (see Hadzhilazova and Mladenov (2010)) and the Sturmian spiral (Mladenov et al. (2011a)). On the other side, Serret states that the curve S_1 (see Fig. 2.5) coincides with Bernoulli's lemniscate, but looking at the figure, one can see that, besides the lemniscate, there exists an extra part of the curve.

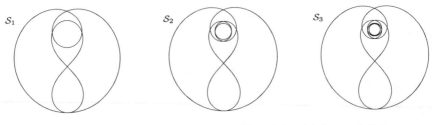

Fig. 2.5 The classical Serret curves S_1 - *left*, S_2 - *middle* and S_3 - *right* for $n = 1, 2, 3$

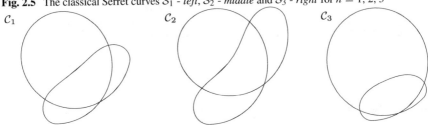

Fig. 2.6 Three examples of the generalized Serret's *curves* C_1 (*left*), C_2 (*middle*) and C_3 (*right*) generated, respectively, with parameter sets $\lambda = 1/3, \sigma = 7/5$, $\lambda = 4/3, \sigma = 9/7$ and $\lambda = 1/7, \sigma = 5/3$

This discrepancy also suggests the necessity of a deeper study of the whole family of Serret curves.

2.9 Cassinian Ovals

This remarkable plane curve is defined as the geometrical locus of the points in the plane for which the product of the distances from two fixed points $\mathbf{F}_1$ and $\mathbf{F}_2$ is a constant, which will be denoted by c^2, and *the distance* $(\mathbf{F}_1, \mathbf{F}_2)$ between $\mathbf{F}_1$ and $\mathbf{F}_2$ is also a constant, denoted by $2a$. In the XOZ plane, the Cassinian ovals are given by the equation (an alternative form is (2.10))

$$(x^2 + z^2 + a^2)^2 - 4a^2x^2 = c^4. \tag{2.74}$$

It is clear that these curves are symmetrical with respect to both coordinate axes. Their shapes depend on the precise relationship of the geometrical parameters a and c. From now on, we will consider the case in which $a < c < a\sqrt{2}$ (this is case *3* in Fig. 2.7). For $c \geqq a\sqrt{2}$, we have ellipse-like figures illustrated by curves numbered as *4* and *5*, and when $c = a$, the curve is given by the equation

$$(x^2 + z^2)^2 = 2a^2(x^2 - z^2), \tag{2.75}$$

Fig. 2.7 Cassinian ovals
drawn with different values
of dimensionless ratio
$\varepsilon = a/c$

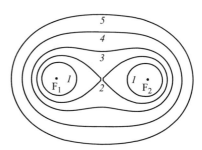

which is nothing less than the Bernoullian lemniscate reproduced here as curve 2.
Finally, in the case in which $a > c$, the curves reduce to two disjoint ovals (these
ovals are depicted as curves *1* in Fig. 2.7).

Cassini has proposed the fourth degree curves (2.74) in an attempt to describe the
planetary motions in the solar system properly. Equations (2.74) and (2.75) describe
concrete algebraic curves. The meaning of the last notion is that the rectangular
coordinates x, z of the points on the curve C in the plane satisfy an algebraic equation

$$F(x, z) = 0, \tag{2.76}$$

where $F(x, z)$ is a polynomial function in its variables.

Following the tradition established by Canham (1970), Deuling and Helfrich
(1976), Funaki (1955) and Vayo (1983), the Cassinian ovals can be considered as a
model of red blood cells. For more detail, see Angelov and Mladenov (2000) and
Hadzhilazova et al. (2011).

2.9.1 Alternative Parameterizations

By making use of the general formula for the curvature of curves defined implicitly
as $F(x, z) = 0$, i.e.,

$$\kappa(x, z) = \frac{|F_{xx}F_z^2 - 2F_{xz}F_xF_z + F_{zz}F_x^2|}{(F_x^2 + F_z^2)^{3/2}}|_{F=0}, \tag{2.77}$$

one can easily find that the curvature of the Cassinian ovals (2.74) is given by the
formula

$$\kappa(r) = \frac{a^4 - c^4}{2c^2r^3} + \frac{3r}{2c^2}. \tag{2.78}$$

This formula makes clear the fact that the curvature is a function depending only
on the polar radius r (as the parameters a and c are considered to be fixed).

2.9.1.1   Integration

Here, we will present a scheme (see Hadzhilazova et al. (2011)), by which one can reconstruct (in principle) any plane curves whose curvature depends solely on the distance from the origin, i.e., $\kappa \equiv \kappa(r)$. For this purpose, let us recall the general formula for the curvature in polar coordinates (see (1.16))

$$\kappa = \frac{r^2 + 2\dot{r}^2 - r\ddot{r}}{(r^2 + \dot{r}^2)^{3/2}}, \tag{2.79}$$

where the dots over r denote the derivatives with respect to the polar angle ϕ. The introduction of the new variable

$$\dot{r} = \frac{dr}{d\phi} = \tau \tag{2.80}$$

has, as a result, the useful formula

$$\phi = \int \frac{dr}{\tau}. \tag{2.81}$$

We also have

$$\ddot{r} = \frac{d^2 r}{d\phi^2} = \frac{d\dot{r}}{d\phi} = \frac{d\tau}{d\phi} = \frac{d\tau}{dr}\frac{dr}{d\phi} = \tau\frac{d\tau}{dr}, \tag{2.82}$$

and therefore

$$\kappa = \frac{r^2 + 2\tau^2 - r\tau\frac{d\tau}{dr}}{(r^2 + \tau^2)^{3/2}} = \frac{2}{(r^2 + \tau^2)^{1/2}} - \frac{r^2 + r\tau\frac{d\tau}{dr}}{(r^2 + \tau^2)^{3/2}}. \tag{2.83}$$

The above formulas also suggest introduction of the notation

$$r^2 = \xi, \qquad r^2 + \tau^2 = \zeta \tag{2.84}$$

and, making use of them, the possibility of writing

$$r^2 + r\tau\frac{d\tau}{dr} = \xi\frac{d\zeta}{d\xi}, \tag{2.85}$$

which ultimately leads to the formula

$$\kappa(\xi) = \frac{2}{\sqrt{\zeta}} - \frac{\xi}{\sqrt{\zeta^3}}\frac{d\zeta}{d\xi} = 2\frac{d}{d\xi}\left(\frac{\xi}{\sqrt{\zeta}}\right). \tag{2.86}$$

The integration of the last equation gives us

$$\frac{\xi}{\sqrt{\zeta}} = \frac{1}{2} \int \kappa(\xi) d\xi. \tag{2.87}$$

Going back to the original coordinates, one ends up with the equation

$$\frac{r^2}{\sqrt{r^2 + \tau^2}} = \int r\kappa(r) dr. \tag{2.88}$$

Performing the integration on the right-hand side produces

$$\int r\kappa(r) dr = m(r) - \omega, \tag{2.89}$$

where ω denotes the integration constant. Solving the system of Eqs. (2.88) and (2.89) for τ, one gets

$$\tau = \frac{r\sqrt{r^2 - (m(r) - \omega)^2}}{m(r) - \omega} \tag{2.90}$$

and, respectively,

$$\phi = \int \frac{m(r) - \omega}{r\sqrt{r^2 - (m(r) - \omega)^2}} dr, \tag{2.91}$$

which is the result that will be used extensively in what follows. Here, the integration constant is omitted, as it is responsible only for the choice of the polar axis, which can be done arbitrarily.

In order to obtain concrete results, one has to specify the curvature in explicit form, which we will do below.

2.9.1.2 Parameterization

The result for the Cassinian oval (2.78), suggests consideration of the curves whose curvature is given by the general formula

$$\kappa = \frac{\lambda}{r^3} + \mu r = \frac{\lambda}{r^3} + 3\nu r, \qquad \lambda \in \mathbb{R}, \quad \nu \in \mathbb{R}^+. \tag{2.92}$$

In what follows, we will present the parametric equations of the curves whose curvature is specified in (2.92) using the method just described. It is easy to see that strongly positive values of the parameters λ and μ exactly reproduce the Cassinian oval due to the relations

$$a^4 = \frac{4\lambda\mu + 1}{4\mu^2}, \qquad c^2 = \frac{1}{2\mu}. \tag{2.93}$$

The above range of parameters could be easily extended by adding negative values of λ which fulfill, together with μ, the inequality

$$4\lambda\mu + 1 > 0. \tag{2.94}$$

Further on, this will be taken for granted, but let us mention that it also includes some negative values of λ, which means that the corresponding curves should actually be considered as a deformation (Mladenov et al. (2011)) of the parent curve (2.78).

Applying formulas (2.89) and (2.91) with (2.92) produces, respectively,

$$m(r) = -\frac{\lambda}{r} + \nu r^3 \tag{2.95}$$

and

$$\phi = \nu \int \frac{r^3 dr}{\sqrt{-\nu^2 r^8 + (2\lambda\nu + 1)r^4 - \lambda^2}} - \lambda \int \frac{dr}{r\sqrt{-\nu^2 r^8 + (2\lambda\nu + 1)r^4 - \lambda^2}}.$$

The above integrals can be uniformized by the following chain of substitutions:

$$r^2 = \chi = n\,\mathrm{dn}(u, k), \qquad n = \frac{\sqrt{2\lambda\nu + 1 + \sqrt{4\lambda\nu + 1}}}{\sqrt{2\nu}}, \tag{2.96}$$

where $\mathrm{dn}(u, k)$ is one of the Jacobian elliptic functions, u is the uniformizing parameter and

$$k = \sqrt{\frac{2\sqrt{4\lambda\nu + 1}}{2\lambda\nu + 1 + \sqrt{4\lambda\nu + 1}}}$$

is the elliptic modulus. As a result, the integrals for the polar angle transform into

$$\phi = \frac{1}{2}\left(\frac{\lambda}{\nu n^2 \tilde{k}} \int \frac{du}{\mathrm{dn}(u, k)} - \int \mathrm{dn}(u, k)du\right), \qquad \tilde{k} = \sqrt{\frac{2\lambda\nu + 1 - \sqrt{4\lambda\nu + 1}}{2\lambda\nu + 1 + \sqrt{4\lambda\nu + 1}}}$$

and can then be evaluated, i.e.,

$$\phi(u) = \frac{1}{2}\left(\frac{\lambda}{\nu n^2 \tilde{k}}\arccos\frac{\mathrm{cn}(u, k)}{\mathrm{dn}(u, k)} - \mathrm{am}(u, k)\right), \tag{2.97}$$

where $\mathrm{am}(u, k)$ is the Jacobian amplitude function and $\mathrm{cn}(u, k) = \cos\mathrm{am}(u, k)$.

Alternatively, the integrals defining the polar angle ϕ can be evaluated via the substitution $r^4 = t$, and as a result, this gives us

$$\phi(r) = \frac{1}{4}\left(\arcsin\frac{2\nu^2 r^4 - 2\lambda\nu - 1}{\sqrt{4\lambda\nu + 1}} + \arcsin\frac{2\lambda^2 - (2\lambda\nu + 1)r^4}{\sqrt{4\lambda\nu + 1}\,r^4}\right). \tag{2.98}$$

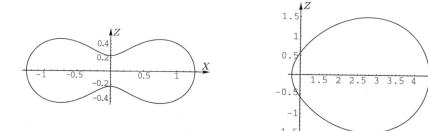

Fig. 2.8 The biconcave *curve* on the *left-hand side* is produced via (2.97) with parameters $\lambda = -0.05$ and $\nu = 0.6$. The internal *oval* on the *right-hand side* is generated via (2.98) with $\lambda = -0.59$ and $\nu = 0.05$

Remark Let us point out that the above parameterizations are entirely different from those reported in Mladenov (2000), Mladenov et al. (2011) and Angelov and Mladenov (2000). Besides, one should note a subtle difference between the formulas (2.97) and (2.98)– the first one is capable of producing both external and internal Cassinian ovals, but not the Bernoullian lemniscate (see Hadzhilazova and Mladenov (2010)), while the second one can be used to draw the lemniscate and the internal ovals, but omits the biconcave curves (see Figs. 2.7 and 2.8).

References

B. Angelov, I. Mladenov, On the geometry of red blood cells. Geom. Integr. Quant. **1**, 27–46 (2000)

M. Berger, B. Gostiaux, *Differential Geometry: Manifolds Curves and Surfaces* (Springer, New York, 1988)

B. Canham, The minimum energy of bending as a possible explanation of biconcave shape of the human red blood cell. J. Theoret. Biol. **26**, 61–81 (1970)

H. Deuling, W. Helfrich, Red blood cell shapes as explained on the basis of curvature elasticity. Biophys. J. **16**, 861–868 (1976)

P. Djondjorov, M. Hadzhilazova, I. Mladenov, V. Vassilev, A note on the passage from the free to the elastica with a tension. Geom. Integr. Quant. **10**, 175–182 (2009)

P. Djondjorov, V. Vassilev, I. Mladenov, Plane curves associated with integrable dynamical systems of the Frenet-Serret type, in *9th International Workshop on Complex Structures, Integrability and Vector Fields*, ed. by K. Sekigawa, V. Gerdjikov, S. Dimiev (World Scientific, Singapore, 2009a), pp. 56–62

C. Doss-Bachelet, J.-P. Françoise, C. Piquet, *Géométrie différentielle* (Ellipses, Paris, 2000)

H. Funaki, Contribution on the shapes of red blood corpuscles. Japan J. Physiol. **5**, 81–92 (1955)

A. Gray, *Modern Differential Geometry of Curves and Surfaces with Mathematica®*, 2nd edn. (CRC Press, Boca Raton, 1998)

M. Hadzhilazova, I. Mladenov, On Bernoulli's lemniscate and co-lemniscate. CRAS (Sofia) **63**, 843–848 (2010)

M. Hadzhilazova, I. Mladenov, Once more the mylar balloon. CRAS (Sofia) **61**, 847–856 (2008)

M. Hadzhilazova, I. Mladenov, P. Djondjorov, V. Vassilev, New parameterizations of the Cassinian ovals. Geom. Integr. Quant. **12**, 164–170 (2011)

E. Jahnke, F. Emde, F. Lösch, *Tafeln Höherer Funktionen* (Teubner, Stuttgart, 1960)

G. Krohs, Die Serret'schen Kurven sind die einzigen algebraischen vom Geschlecht Null, deren Koordinaten eindeutige doppelperiodische Functionen des Bogens der Kurve sind. Inaugural Dissertation Halle-Wittenberg Universität 74 p. (1891)

J. Liouville, A note on the Serret's article. J. Mathematiques Pures et Appliquées **10**, 293–296 (1845)

A. Lipkovski, Serret's curves, in *Book of Abstracts of the International Topological Conference Topology and Applications*, 191–192 Phasis, Moscow (1996)

S. Matsutani, Euler's elastica and beyond. J. Geom. Symmetry Phys. **17**, 45–86 (2010)

I. Mladenov, Uniformization of the Cassinian oval. CRAS (Sofia) **53**, 13–16 (2000)

I. Mladenov, M. Hadzhilazova, P. Djondjorov, V. Vassilev, On some deformations of the Cassinian oval, in *AIP Conference Proceedings* vol 1340, (New York 2011) pp 81–89

I. Mladenov, M. Hadzhilazova, P. Djondjorov, V. Vassilev, On the plane curves whose curvature depends on the distance from the origin, in *AIP Conference Proceedings*, vol. 1307 (New York, 2010), pp. 112–118

I. Mladenov, J. Oprea, Balloons, domes and geometry. J. Geom. Symmetry Phys. **15**, 53–88 (2009)

I. Mladenov, M. Hadzhilazova, P. Djondjorov, V. Vassilev, On the generalized Sturmian spirals. C. R. Bulg. Acad. Sci. **64**, 633–640 (2011a)

I. Mladenov, M. Hadzhilazova, P. Djondjorov, V. Vassilev, Serret's curves, their generalization and explicit parametrization, in *Geometric Methods in Physics*, ed. by P. Kielanowski, S.T. Ali, A. Odzijewicz, M. Schlichenmaier, Th Voronov (Basel, Birkhäuser, 2012), pp. 383–390

F. Olver, D. Lozier, R. Boisvert, Ch. Clark (eds.), *NIST Handbook of Mathematical Functions* (Cambridge Univ. Press, Cambridge, 2010)

J. Oprea, *Differential Geometry and Its Applications*, 3rd edn. (Mathematical Association of America, Washington D. C, 2007)

V. Rovenski, *Geometry of Curves and Surfaces with Maple*® (Birkhäuser, Boston, 2000)

J.-A. Serret, Sur la représentation géometrique des fonctions elliptiques et ultra-elliptiques. J. Mathematiques Pures et Appliquées **10**, 257–290 (1845)

J.-A. Serret, Sur les courbes elliptiques de la première classe. J. Mathematiques Pures et Appliquées **10**, 421–429 (1845a)

D. Singer, Curves whose curvature depends on distance from the origin. Am. Math. Monthly **106**, 835–841 (1999)

V. Vassilev, P. Djonjorov, I. Mladenov, Integrable dynamical systems of the Frenet-Serret type, in *Proceeding 9th International Workshop on Complex Structures, Integrability and Vector Fields*, ed. by K. Sekigawa, V. Gerdjikov, S. Dimiev (World Scientific, Singapore, 2009) pp. 234–244

H. Vayo, Some red blood cell geometry. Can. J. Physiol. Pharmacol. **61**, 646–649 (1983)

R. Yates, *Curves and Their Properties* (Edwards Brothers, Ann Arbor, 1959)

C. Zwikker, *The Advanced Geometry of Plane Curves and Their Applications* (Dover, New York, 1963)

Chapter 3
Biological Membranes

Abstract This chapter plays the role of an introduction to the subject of membranology. Here, we have gathered information and definitions regarding biological membranes that will subsequently become necessary. The principal purpose is to convince the reader that the available information on various origins is sufficient to reach the fully reasonable conclusion that the membranes can be viewed as two-dimensional surfaces in the three-dimensional Euclidean space and that the fundamental role in the shape of the membranes is played by the bending energy. The latter, in turn, is directly related to the membrane's curvatures. This chapter surveys a series of biological, physical and mechanical aspects that are fundamental as a foundation for our results. This chapter plays a supporting role for our further considerations. We have collected required notations and definitions related to biological membranes, which we will make use of later. For more information, the reader can see the cited literature.

3.1 Subject Matter and Biological Membranes

All organisms are made of the building blocks known as cells. However, some forms of life are made of only a single cell, such as the many species of bacteria and protozoa, while the others consist of vast numbers of cells—these are called multicellular.

The cell has all characteristics of the living systems: the ability to exchange substances and energy with the environment (and as a result, to grow, multiply and hand down hereditary information), to respond to external stimuli and to move about as needed. A cell is the smallest structural and functional unit of life.

The existing cells have different shapes. Many bacterial and unicellular organisms are spherical in shape. Animal epithelial cells are usually polyhedral. The spindle form was established in smooth muscle and extremely stretched plant fibers. The vascular cellular tissues of plants are tubular.

The term 'membrane' is used to denote the cell boundary, which is the barrier between the cell and the environment on one side, and the selective barrier that allows some things to pass through but stops others on the other side. The name

© Springer International Publishing AG 2017
I.M. Mladenov and M. Hadzhilazova, *The Many Faces of Elastica*,
Forum for Interdisciplinary Mathematics 3, DOI 10.1007/978-3-319-61244-7_3

plasma membrane was introduced by Nägeli (1855), when he was considering the penetration of pigments into plant cells.

Biomembranes are complex organized macromolecular structures of the cells, often consisting of a phospholipid bilayer with embedded, integral and peripheral proteins used in communication and the transportation of chemicals and ions.

Biological membranes support the spacial organization of life. The membranes define the boundaries of the living cells and protect the cell from its surroundings.

Membranes are not static barriers: rather, they are active structures. They are selectively permeable to ions and organic molecules, control the movement of substances in and out of the cells, and represent the main structural element of the cell. Biomembranes create opportunities for decreasing diffusion and increasing the local concentration of metabolites. Therefore, their metabolite functions are higher speed and greater efficiency, and many of the reactions cannot run in a homogeneous water medium, even if all necessary enzymes and substances exist there. Biomembranes have a relatively large surface area, due to inter-cellular exchange.

The structure and dynamical properties of biomembranes are studied through a number of methods for determination of chemical composition and physical properties of the components of either natural and or artificial and reconstructed membranes. Such methods include: electron microscopy, X-ray structural analysis, spectroscopy (visible, ultraviolet, infrared, EPR, NMR), fluorescence, polarimetry, light scattering. A more detailed description of these methods and techniques can be found in Evans and Skalak (1980), Marinov (2001) and Lipowsky and Sackmann (1995).

3.2 Types of Membranes

Membranes are divided into natural (biomembrane), artificial (model) and reconstructed membranes. All cell membrane shapes have inner and outer surfaces, with different contents, structures and functions. Due to differences in content, the structure and functions of the inner and outer surface of the membranes are divided into symmetric and antisymmetric. In artificial membranes, the antisymmetry is less obviously expressed.

Isolation of the membranes of the cells is usually accompanied by changes in those membranes. Native membranes are membranes that are fully preserved structurally and functionally; while other membranes have been changed, they are called modified. The study of the modified membrane is important for clarification of the pathology of the cells.

The reconstructed membranes are formed through the self-organization of fragments from natural membranes or by lipids and proteins that have been isolated from them. They take place between natural and artificial membrane. The chemical structure of artificial membranes consists of lipids, proteins or other substances. According to the manner of their production, they have a different structure and are divided into two main groups—flat and spherical, which can be monolayer, bilayer or multilayer. The flat monolayer is formed at the water/organic solvent or water/air border from a given molecular row, oriented with their hydrophilic parts to the water phase. The monolayer is the same as one half of the natural membrane. The simplest model of biomembrane is the flat bilayer membrane, which is made up of two surfaces contacting with the water. These are called "black" membranes. The term 'black bilayer' refers to the fact that they are dark in reflected light because the thickness of the membrane is only a few nanometers, so the light reflecting off the back face destructively interferes with light reflecting off the front face. Spherical membranes are formed from lipids dispersed in a water medium by different methods. Their structure and size vary over wide ranges. The micelles are the smallest entities constructed from a single layer of lipids, the hydrophobic parts of which are located in the center. Spherical membranes composed of one or more bilayers are called liposomes. Their size usually varies is from 25 nm to 1 μm, but it can be even larger. The large variety of artificial membrane presents the opportunity for research, and this helps us to find the structure and properties of the natural membranes. That is why they are defined as model membranes.

3.3 Functions of Biomembranes

Biomembranes are highly organized structures, consisting mainly of phospholipid bilayers with embedded, integral and peripheral proteins. About 80% of the dry weight of most cells consists of membranes. The main function of the biological membrane is its role as the best barrier between aquatic surrounding and the hydrophobic layer. The water-soluble compounds, which are in the cells and their surroundings, are not soluble in the lipid medium of the membrane and either pass very slowly or do not pass at all through very thin lipid layers. The structure is flexible and helps with growth and movement, as well as with action of the protein structure. And finally, the structures have a low dielectric constant, which gives electrical properties to the membrane that it uses for its functions. The main membrane functions are:

(a) Containment and separation
Membranes separate cytoplasms from their surroundings, which determines the personality of each cell. Eukaryotic cells are structurally complex, and, by definition, are organized, in part, by interior compartments that are themselves enclosed by lipid membranes that resemble the outermost cell membrane. The compartments, together with their membrane, form the cytoplasm organelles: nucleus, mitochondria, lysosome, chloroplast, etc. The endoplasmic reticulum also belonging to the intracellular membrane.

(b) Transport of substances
The membrane is a barrier with highly selective penetration, which has specific molecular pumps and channels. These are characterized by high speed and very high efficiency, as biomembranes have a large surface. The free energy defines the movement of molecules and ions through the membrane in response to concentration gradients and membrane potentials.

(c) Energy conversion

In some membrane systems, which contain highly organized sets of enzymes and other proteins, the most important processes of energy conversion in biological systems take place. Photosynthesis is a process used by plants and other organisms to convert light energy, normally from the Sun, into chemical energy that can later be released as fuel for the organisms' activities (energy transformation), and oxidative phosphorylation—synthesis of ATP at the expense of oxidation of organic substrates. The first process runs into the inter-membrane of the chloroplast and the second in the inter-membrane of the mitochondria.

(d) Recognition of molecules and cells, transmission and generation of signals

Most animal epithelial cells have a fuzz-like coat on the external surface of their plasma membranes with a thickness up to 100 nm, which is called the glycocalyx. This coating consists of several carbohydrate moieties of membrane glycolipids and glycoproteins, which serve as backbone molecules for support. Generally, the carbohydrate portion of the glycolipids found on the surface of plasma membranes helps these molecules contribute to cell-to-cell recognition, communication, and intercellular adhesion.

3.4 Chemical Composition and Physical Properties of Biomembranes

Every biomembrane has specific content due to its cytological, tissue and organism type. The main components of biomembranes are lipids and proteins. The content of the proteins is from 20 to 85%, depending on the kind of membrane. The membranes contain about 10–15% carbohydrates connected to protein molecules (glycoproteins) or lipids (glycolipids). For every membrane, the content and the main types of lipid and protein are specific, permanent and genetically determined. It changes within definite boundaries depending on the ages, functions and metabolic status and according to the pathologies.

The water in membranes is about 20–30% of their fresh weight. The main part of it forms a hydrated shell around the polar parts of the lipid and protein molecules. This is not osmotic active water and it cannot solve any substances. Besides this water, there is free water, and sometimes there is connected water between lipid layers.

The basic physical parameters of biomembranes are: thickness, elasticity modulus, surface tension, index of refraction, electric capacity, resistance and voltage for membrane electroporation, permeability for different substances and their energy of activation, transmembrane and electrokinetic potential, and electric charge.

The thickness of the biomembrane is about 11 nm by measurement of small angles through the scattering of X-rays and about 7 nm through electron microscopy. All biomembranes have a three-layered image in electron microscopy, consisting of two dark layers of about 2 nm with a light strip about 3–3.5 nm between them.

Fig. 3.1 Structure of lipid molecule

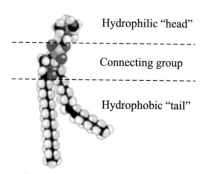

Hydrophilic "head"

Connecting group

Hydrophobic "tail"

Phosphatidylcholine

Considering the mechanical properties of the biomembranes with respect to their molecular structure, we have in mind the following assumptions:

(a) the thickness of the membrane does not change as the result of mechanical impacts.
(b) the distance and location of the membrane molecules are determined by electrostatic and van der Waals forces.
(c) the different membrane molecules, especially the lipids, can very safely be assumed to be cylinders or cones. The membrane segment, composed of such conical molecules, will be bended even without any external tension on it.

3.4.1 Molecular Structure and Physicochemical Properties of Membrane Lipids

Lipids (from 'lipos', meaning 'fat' in Greek) are the main structural components of membranes. They are a group of compounds with the same physicochemical properties, more precisely, with limited solubility in water and polar solvents, but with high solubility in non-polar solvents (chloroform, ether, benzene). Lipids have a huge variety of chemical structures. The structure of the lipid molecule is shown in Fig. 3.1.

One end of the lipid molecule consists of hydrophilic (soluble in water) groups that form the polar "head". The other end is made of a long hydrocarbon chain that is insoluble in water is called a hydrophobic "tail". The molecules of this type that simultaneously have hydrophilic and hydrophobic parts are called amphiphilic molecules. Amphiphilic molecules have a tendency towards aggregation.

The connecting group between the hydrophilic and hydrophobic parts of the molecule define the name of the lipid in the primary classification. Usually, it is fatty polyvalent alcohol, which contains two or three hydroxyl groups. For example, for glycerol, the lipid component consists of a group of glycerolipids. More than half of the lipids in the nature are glycerides. If the connecting group is sphingosine, the lipid is sphingolipid. The richest of the sphingolipids are the nerve tissues, especially the brain.

The polar "heads" of all lipid molecules are negatively charged or electrically neutral. This is important, as the total negative electric charge of the membrane plays an important role in the behavior of the cells. The forces which keep the lipid molecules in aggregation are the hydrophobic forces between the molecule tails. Electrostatic forces between their heads have an impact on them too. Monolayers containing lipids with saturated and unsaturated fatty acid residues are more liquid than layers containing only saturated residues.

Cholesterol influences the binding and mobility of the lipid molecules within membranes. If we add cholesterol to the lipid monolayer in small portions, the area of the monolayer decreases instead of increasing. This continues until the concentration of cholesterol becomes so high that, for every two lipid molecules, there is one cholesterol molecule. Over this ratio, the area of the monolayer starts to increase. The explanation for the decrease in the area is that one molecule of cholesterol enters between two lipid molecules. This straightens their hydrophobic tails and gives these molecules the opportunity to move closer to one another.

The cholesterol is embedded in the membrane with the same orientation as the phospholipid molecule. The hydroxy group of the cholesterol interacts with the polar head groups of the membrane phospholipids and sphingolipids, while the bulky steroid and the hydrocarbon chain are embedded in the membrane, alongside the non-polar fatty-acid chain of the other lipids. Through the interaction with the phospholipid fatty-acid chains, the cholesterol increases membrane packing. For a large amount of cholesterol the area of the monolayer grows at the expense of the area of "redundant" for packing cholesterol molecules.

3.4.2 Membrane Proteins and Glycoproteins

The proteins perform all specific functions of the membranes. The determination of membrane protein structures has remained a challenge, in large part due to the difficulty of establishing experimental conditions in which the correct conformation of the protein in isolation from its native environment is preserved.

Chemically, the proteins are divided into simple and complex. The simple ones are hydrolyzed by acids or bases to amino acids. Besides simple proteins in the plasma membranes of all eukaryotic cells, there are complex proteins which also contain a carbohydrate moiety, e.g., these are glycoproteins and proteoglycans in which the carbohydrate moiety can vary from less than 1% to more than 30%.

3.4.2.1 Membrane Proteins Functions

According to their functions in the membrane, proteins are divided into enzymes, transport proteins, receptors and structural proteins.

The enzymes may have many activities, such as oxidoreductase, transferase or hydrolase. The main part of membrane proteins are enzymes, and from the variety of enzyme activity, we can derive conclusions about the protein content within it.

The transport protein is a membrane protein involved in the movement of ions, small molecules, or macromolecules, such as another protein, across a biological membrane. For example, the transmembrane protein Na^+/K^+ pump (sodium-potassium pump), which is an enzyme, is a solute pump that pumps sodium out of cells while pumping potassium into them, both against their concentration gradients. This pumping is active (it uses energy from ATP) and is very important for cell physiology.

Receptors are specialized integral membrane proteins that allow for communication between the cell and the world outside. The extracellular molecules may be hormones, neurotransmitters, cytokineses, growth factors, cell adhesion molecules, or nutrients. All of them react with the receptor to induce changes in the metabolism and activity of a cell.

Structural proteins are highly hydrophobic and connect with each other spontaneously, although with weak connections. Spectrin is a typical example of structural protein. It lines the intracellular side of the plasma membrane in eukaryotic cells. Spectrin forms pentagonal or hexagonal arrangements, creating a scaffolding and playing an important role in the maintenance of plasma membrane integrity and cytoskeletal structure.

3.4.2.2 Localization and Connection of the Proteins in a Membrane

According to their location and association in the membrane the proteins are, respectively, peripheral membrane proteins and integral membrane proteins. Peripheral membrane proteins are temporarily attached either to the lipid bilayer or to the integral proteins by a combination of hydrophobic, electrostatic, and other non-covalent interactions. Peripheral proteins dissociate following treatment with a polar reagent, such as a solution with an elevated pH or high salt concentrations.

Integral membrane proteins are permanently attached to the membrane. Such proteins can be separated from the biological membranes only by using detergents, non-polar solvents, or sometimes denaturing agents. Figure 3.3 depicts transmembrane proteins. Proteins which cross the membrane only once are "single-pass" transmembrane proteins, while "multi-pass" membrane proteins weave in and out, crossing the membrane several times.

3.5 Membrane Models and Methods for the Study of Biomembranes

3.5.1 Modern Theories

A long time ago, Singer and Nicolson (1972) proposed the "fluid mosaic" model of the membrane (see Fig. 3.2).

According to this model, the basis of the biomembrane is a lipid bilayer in which hydrocarbon chains of the phospholipid molecules are in the liquid crystal phase. Proteins were visualized as macromolecules embedded in the bilayer in an iceberg-like fashion, penetrating either half or all the way through (see Fig. 3.2). The protein molecules were visualized as being completely free to translate laterally in the liquid bilayer. Besides the fact that the fluid mosaic model is widely accepted, it presents a complex and varied system (such as the biological membrane) which is simplified and schematic. One of the major postulates of this model is the assumption of free movement of lipid and protein molecules. Later on, became clear that not all lipids and proteins can be moved freely, and that in some cases, their mobility is highly limited. In many membranes, the integrated proteins are fixed due to a high concentration or aggregation of proteins, resulting from formation of lipid domains around them and interactions with the cytoskeleton, formed by the internal structure of the cell.

In certain membrane lipids, distribution on their surface is not homogeneous, as should be expected in the case of free diffusion. To perform vector functions, cell membranes have to be asymmetric, i.e., the components of bilayers are distributed differently on the outer and inner surfaces. In this asymmetry, either lipids or protein components are involved. Lipid asymmetry is relative, so the same lipids may be present in the outer and inner monolayers, but with a different concentration. According to the Singer and Nicolson model, the asymmetric location of lipids is preserved if their molecules pass slowly enough from one side of the membrane to the other. However, the newer studies shows that this can be done rather quickly (with a half period of 1–2 min).

Figure 3.3 presents an up-to-date model of a biological membrane in which the integral proteins can have a globular structure as well as conformation of an α—helix.

These are glycoproteins, and their oligosaccharide chains, as the glycolipid's are always outside of the membrane. This position and the one-way action of some mem-

Fig. 3.2 Fluid mosaic model
of Singer and Nicolson

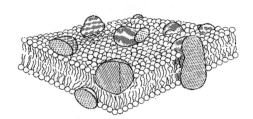

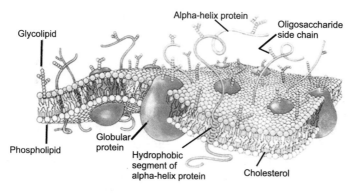

Fig. 3.3 Modern model of biological membrane

brane transport systems (pumps) is another manifestation of membrane asymmetry. It leads to functional differences between both surfaces of the membrane.

Two main factors can influence a membrane's asymmetry. The first is metabolism, since the newly synthesized molecules entering constantly into the membrane, and the part of the present degrade, becoming the two sides of the membrane at different speeds. The second factor is the different physical interactions (electrostatic, van der Waals, etc.) of membrane components with each other and with their environment, which may lead to asymmetric distribution of the lipid molecules.

Besides, proteins in the phospholipid layer are submerged and molecules of cholesterol, which are rigid and shorter than the lipid molecules. They maintain their polar heads as relatively fixed and sorted, while the more remote parts of their tails can freely alter their position.

3.6 Model Membrane Structures

3.6.1 Lipid Associates

In aqueous solutions and on the water/air boundary the lipids form aggregates. The molecules in these aggregates are connected to van der Waals forces. The most popular associates are monolayers, micelles and bilayers.

3.6.1.1 Monolayers

Monolayers are formed on the water/air or water/oil boundary, after a solution of lipid in a volatile solvent is dripped on the water's surface. After the solvent evaporates we obtain a cover with a thickness approximately the length of the lipid molecules, in which the polar heads of the molecules are turned into water and their hydrocarbon chains—into air. When there are no longer any restrictions, the lipid cover occupies the maximum possible area and becomes a system analogous to the two-dimensional gas.

When reducing the monolayer gradually, its surface tension increases, which can be defined as the difference between surface tension of the pure water and the surface tension of the water with a monolayer.

If the monolayer's density increases, its molecules start to interact, and a continuous lipid film is subsequently formed on the water surface in the form of a liquid monolayer or a two-dimensional liquid. Further contraction of the monolayer leads to the dense packing of lipid molecules where polar heads come up to each other and hydrocarbon chains rise in the air but preserve a certain mobility. This condition is called a condensed monolayer. If the surface pressure increases further, the condensed monolayer becames hard, practically incompressible, and every single lipid molecule takes up minimal area. The cover starts to disintegrate after exceeding a boundary value of the pressure, because the molecules of the monolayer come up to one another (a state of collapse).

Some macromolecules have significant influence on the properties of the monolayer, e.g., proteins added into the water. Through the change in area occupied by one molecule, the surface tension and the electric potential of the monolayer, we can explore the interaction between proteins and lipids. Lipid monolayers are the first models of biological membranes.

3.6.1.2 Micelles

Micelles are the simplest aggregates that the lipid molecules can form in the bulk solvent phase—(Fig. 3.4a). Depending on the solvent, a micelle can be normal or inverse.

In normal-phase micelle, the head groups of the lipid molecules are toward the water phase, and the tails form, hydrophobic core isolated from the water surroundings (oil-in-water micelle). Inverse micelles have the head groups at the centre with the tails extending out to the solvent (water-in-oil micelle).

The Lipid molecular construction, and especially the size of the polar and nonpolar parts of the molecule, are essential to the lipid's ability to form micelles. In water, they easily form micelles lipids which have huge polar heads and/or relatively small hydrocarbon chains (the molecules have the form of an inverted cone). The small volume of the polar heads, the neutralization of their charges and the massive hydrocarbon chains (the molecules have the form of a cone) help in the formation of reverse micelles in nonpolar solvents.

Phospholipids which cannot form micelles in a water medium (e.g., phosphatidylcholin) may form them after mixing with surfactants.

Fig. 3.4 Structures that can
be formed by phospholipids
in aqueous solutions: **a**
micelle, **b** liposome and **c**
lipid bilayer

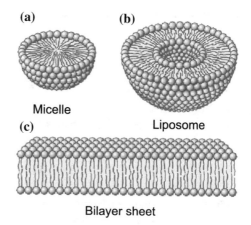

(a)   **(b)**

Micelle

(c)   Liposome

Bilayer sheet

3.6.1.3   Bilayers

For the lipids which cannot form micelles in water, thermodynamically, the most
favorable association form is the bilayer. The lipids that most easily form bilayers
are those for which the area occupied by the polar head and the cross-section of
hydrocarbon chains differ slightly, i.e., the molecule has the form of a cylinder. This
is the case for the most phospholipids in biomembranes composition. The aggregated
lipid molecules in the bilayer are situated in two parallel monolayers, facing each
other with their hydrophobic sides. The polar groups form two hydrophilic surfaces
that separate the inner hydrocarbon phase of the bilayer from the water medium (see
Fig. 3.4c).

In the water medium, the phospholipids and the glycolipids easily form bilayers
instead of micelles, which is why they are the main components of the biomembranes.
Bilayers reach macroscopic sizes up to a millimeter, while micelles are generally
small—with a diameter less than 20 nm. Furthermore (although in their liquid state),
the bilayers act as barriers to permeability through them.

The ability of lipid molecules to form bilayers quickly and spontaneously in
water (self-assembling, self-organizing) is due to their structure and is determined
mainly by their amphipathic properties. The main forces that provide this are the
hydrophobic interactions as a result of the displacement action of water on non-polar
compounds. Furthermore, van der Waals forces act between the hydrocarbon chains,
pack them tightly in bilayers. Between the polar heads and the water molecules arise
electrostatic forces and hydrogen bonds, so that in the stabilization of lipid bilayers,
all the forces of intermolecular interactions in biological systems are involved.

Three biologically important properties of lipid bilayers result from these and
other non-covalent interactions:

(a) a tendency to increase their surface
(b) the ability to close and form a bounded domain (compartment), so as not to allow
    contact between the water and the hydrocarbon chains

(c) the ability to self-sealing, as every hole in the bilayer is energetically un-
favourable.

3.6.2   Model Artificial Membranes

3.6.2.1   Bilayer Lipid Membranes

Bilayer lipid membranes are a widely used experimental model, which allows for
reproducing many properties and characteristics of the biomembrane under artificial
conditions. A bilayer lipid membrane is a single lipid bilayer, attached perimetri-
cally to a hole in the inert barrier from the hydrophobic material (polythene, teflon),
separating it into water volumes.

Mueller et al. (1962) first described the formation of such a membrane. The drop
of lipid solution in an organic solvent is applied with a brush under water onto a
hole with a diameter of 1–2 mm in the hydrophobic barrier. Initially, it forms a thick
gray cover, which subsequently becomes thinner, due to an increase in the surface
tension. Then, the color changes to different colors of the spectrum. This is due to the
interference of light reflected from both of its surfaces, meaning that the thickness
of the cover is roughly that of a wavelength of light (600–1000 nm). At last, the
two lipid monolayers form one bilayer with a thickness of 4–6 nm, which reflects
the visible light very slightly and therefore appears black. During formation of the
bilayer the solvent is forced to the edges of the hole, where the layer is expanding,
but it can also be gathered in the middle, where it forms microlenses.

Bilayer membranes that do not contain a solvent (and which are therefore called
dry) can be produced, according to Takagi et al. (1965) and Montal (1974), through
a method involving the mechanical contact of two lipid monolayers, formed on the
water/air boundary. An advantage of this method is the possibility of obtaining asym-
metrical membranes from initial monolayers with different chemical composition.

The instability of the bilayer membranes is due to different factors: the presence
of unwanted impurities, lipid oxidation, unsuitable solvent, temperature fluctuations,
vibrations, rapid changes in concentration and the viscosity of the medium, etc.

However, the lipid bilayer has an extremely high mechanical strength and elasticity. It endures a gentle touch with a glass rod, bending under pressure. It deforms
under mechanical action or osmotic pressure, changing its shape to hemispherical,
and vice versa. The lipid bilayer can be perforated with a hair, a thin wire or a micro
electrode without being destroyed when they are draw out. This demonstrates the
ability of the bilayers to self-adhere.

If we act with a strong electric field for a short time, these membranes have extremely high stability. We have an electric breakdown of the electric field at about
$40\,mV.nm^{-1}$. Their dielectric properties are almost the same as those of the best
known insulators—solid paraffin, porcelain, polyvinylchloride. If there is a breakdown, unlike with solids, the membranes completely recover their functions after
switching off the field. Because of the ability to conduct various electrical measure-

ments (of the electric capacity, conductivity, breakdown voltage, membrane potential, etc.), bilayer membranes have an exclusive and an important role in the study of ion transport across biomembranes.

One new perspective on making bilayer lipid membranes closer to biomembranes is inserting membrane proteins into them. This can be done through the fusion of liposome-containing protein with a lipid bilayer under osmotic stress conditions or under the action of calcium ions and other agents, supporting the fusion. There are real possibilities for creation of biosensor systems based on bilayer membranes.

3.6.2.2 Liposomes

Bangham et al. (1995) have shown that phospholipids in water spontaneously form spherical vesicles called liposomes, which consist of many enclosed lipid bilayers, separated by layers of water (see Fig. 3.4b). Liposomes are used as model membranes in clarifying the number of issues related to the molecular organization and functioning of biomembranes. Depending on their size and the number of lipid bilayers forming them, they are divided into three main groups: (a) multilayer (multilamelar) liposomes, having a diameter of 5–10 μm and a structure comparable to that of a bulb or a cabbage with hundreds of layers; (b) a big monolayer with a diameter of 50–200 nm or bigger; (c) a small monolayer (monolamelar) with a diameter of 20–50 nm.

The ability of phospholipids to form liposomes depend on the temperature of the phase transition. Unsaturated phospholipids that, at room temperature, are in a liquid crystal state easily form liposomes. Those with saturated fatty-acid chains form liposomes only at temperatures that are above the phase transition temperature. The most essential factors are the size of the molecule and the nature of the polar head.

In hypotonic solution, they swell from the water being absorbed, and in hypertonic, they lose water through interlamellar spaces and shrivel. Part of the water in the liposomes remains osmotically inactive. Unlike those of multilayer, small unilamellar liposomes do not exhibit osmotic activity.

The internal water volume of the liposomes can be switched for other substances, which are inserted in advance in the initial lipid-aqueous phase. After that, we can separate the liposomes and suspend them in another solution. In this way one can create conditions for the exchange of substances between the liposomes and their environment through the lipid layer, which is a diffusion barrier. That is why liposomes are widely used for clarifying the lipid barrier function and for the modeling of different transport processes.

Further information in the context of membrane models can be found in books by Lipowsky and Sackmann (1995) and Ou-Yang et al. (1999). The geometrical aspects of these models are treated in-depth by Tu and Ou-Yang (2003), Tu and Ou-Yang (2004) and Tu (2011).

References

D. Bangham, M. Standish, C. Watkins, Diffusion of univalent ions across the lamellae of swollen phospholipids. J. Mol. Biol. **13**, 238–252 (1965)

E. Evans, R. Skalak, *Mechanics and Thermodynamics of Biomembranes* (CRC Press, Boca Raton, 1980)

R. Lipowsky, E. Sackmann, *Handbook of Biological Physics vol. 1: Structure and Dynamics of Membranes* (Elsevier, Amsterdam, 1995)

M. Marinov, *Biophysics (in Bulgarian)* (Medical Univ, Sofia, 2001)

M. Montal, in *Formation of Bimolecular Membranes From Lipid Monolayers*. Methods in Enzymology XXXII. Biomembranes. Part B, eds. By S. Fleischer, L. Pacher (Academic, New York, 1974), pp. 545–554

P. Mueller, O. Rudin, T. Tien, C. Weescott, Reconstitution into an exitable system. Nature **194**, 979–980 (1962)

C. Nägeli, Pflanzenphysiologische Untersuchungen (1855–1858) (1855)

Z.-C. Ou-Yang, J.-X. Liu, Y.-Z. Xie, *Geometric Methods in the Elastic Theory of Membranes in Liquid Crystal Phases* (World Scientific, Hong Kong, 1999)

J. Singer, L. Nicolson, The fluid mosaic model of the structure of cell membranes. Science **175**, 720–731 (1972)

M. Takagi, K. Azuma, U. Kishimoto, A new method for the formation of bilayer in aqueous solution. Ann. Rep. Biol. Works Fac. Sci. Osaka Univ. **13**, 107–110 (1965)

Z.-C. Tu, Z.-C. Ou-Yang, A geometric theory on the elasticity of bio-membranes. J. Phys. A Math. Gen. **37**, 11407–11429 (2004)

Z.-C. Tu, Z.-C. Ou-Yang, Lipid membranes with free edges. Phys. Rev. E **68**, 061915-1-7 (2003)

Z.-C. Tu, Geometry of membranes. J. Geom. Symmetry Phys. **24**, 45–75 (2011)

Chapter 4
Surface Tension and Equilibrium

Abstract In this Chapter, the equilibria of membranes is treated from the perspective of the first principles of mechanics. For example, the Laplace–Young equation, which is valid for an arbitrary infinitesimal area of the membrane, is derived in full detail just to clarify the appearance of membrane curvatures in all further considerations. If we limit ourselves to the class of axially symmetric membranes, the corresponding balance equations can be written in a vector form, which is valid along the whole surface. Projection of this vector equation along the normal and tangential directions at any point leads to a system of two coupled nonlinear equations. It turns out that this system can be solved in two cases of direct interest. Namely, Delaunay surfaces and the polyester balloon, both of which are described explicitly. It should also be noted that the so-obtained parameterization of Delaunay surfaces is regular, contrary to the one which follows from the original construction. Here, we also consider the problem of a fluid body rotating with a constant angular velocity and subjected to surface tension of ambient interface. Determining the equilibrium configuration of this system turns out to be equivalent to the geometrical problem of finding the surface of revolution with a prescribed mean curvature. In the simply connected case, the equilibrium surface can be parameterized explicitly via elliptic integrals of the first and second kinds. Here, we present two different parameterizations of rotating drops based on the Jacobian and Weierstrassian elliptic functions and integrals. By making use of the second of these, we are able to study the finer details of the drop surfaces, such as the existence of closed geodesics and other quantities of geometrical and mechanical interest.

4.1 Mechanical Equilibrium

4.1.1 Laplace–Young Equation

Let us consider the infinitesimal curvilinear quadrangle $ABCD$ of the membrane. If the point A coincides with the origin of the orthogonal coordinate lines on the membrane with $\mathcal{R}_u$ and $\mathcal{R}_v$ being their curvature radii and $p_{\text{in}}, p_{\text{out}}$ being, respectively, the inner and outer pressures, the work W needed for the infinitesimal expansion of

© Springer International Publishing AG 2017
I.M. Mladenov and M. Hadzhilazova, *The Many Faces of Elastica*,
Forum for Interdisciplinary Mathematics 3, DOI 10.1007/978-3-319-61244-7_4

Fig. 4.1 An infinitesimal
local expansion of the
surface of the membrane
under increasing of the
pressure

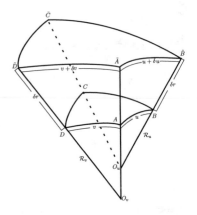

$ABCD$ with sides u and v to the quadrangle $\tilde{A}\tilde{B}\tilde{C}\tilde{D}$ with sides $u + \delta u$ and $v + \delta v$ (δu and δv are their respective infinitesimal increments) is given by the expression

$$W = (p_{\text{in}} - p_{\text{out}})S\delta r = \Delta p S \delta r = \sigma \Delta S$$
$$= \sigma[u(1 + \frac{\delta r}{\mathcal{R}_u})v(1 + \frac{\delta r}{\mathcal{R}_v}) - uv] = \sigma(\frac{1}{\mathcal{R}_u} + \frac{1}{\mathcal{R}_v})uv\delta r = \sigma(\frac{1}{\mathcal{R}_u} + \frac{1}{\mathcal{R}_v})S\delta r,$$

and therefore

$$\Delta p = 2\sigma H, \tag{4.1}$$

where δr is the infinitesimal displacement of the membrane under the pressure difference $\Delta p = p_{\text{in}} - p_{\text{out}}$, σ is the surface tension, and Eq. (4.1) bears the name Laplace–Young equation (Fig. 4.1).

4.2 Tensions and Geometry

4.2.1 Membrane Geometry

Our membrane will be modeled by a surface of revolution about the z-axis generated by a profile curve $(r(s), z(s))$ in the (first quadrant of the) xz-plane (where s is the arclength parameter and we take $z(s)$ to increase with increasing s: $z(s)$ rising from the x-axis and meeting the z-axis orthogonally). This surface has a parameterization

$$\mathbf{x}(s, v) = (r(s) \cos v, r(s) \sin v, z(s)) = r(s)\mathbf{e}_1(v) + z(s)\mathbf{e}_3(v),$$

where the unit radial vector is $\mathbf{e}_1(v) = \cos v\, \mathbf{i} + \sin v\, \mathbf{j}$ and $\mathbf{e}_3(v) = \mathbf{k}$. We also take $\mathbf{e}_2(v) = \mathbf{k} \times \mathbf{e}_1 = -\sin v\, \mathbf{i} + \cos v\, \mathbf{j}$, the unit vector along the parallels of revolution. A *meridian* $r(s)\mathbf{e}_1(\mathring{v}) + z(s)\mathbf{k}$ (i.e., fixed $\mathring{v}$) has the tangent vector $\mathbf{t} = r'(s)\mathbf{e}_1(\mathring{v}) + z'(s)\mathbf{k}$, where the primes denote differentiation with respect to s. Because we parametrize the meridian by arclength, the tangent vector has a unit length, i.e., $r'(s)^2 + z'(s)^2 = 1$. Hence, we can define

$$r'(s) = -\sin\theta(s), \qquad z'(s) = \cos\theta(s), \tag{4.2}$$

where $\theta(s)$ is the angle between $\mathbf{t}$ and $\mathbf{k}$, and write $\mathbf{t} = -\sin\theta\,\mathbf{e}_1 + \cos\theta\,\mathbf{k}$. Note that, since we assume z increases with s, $\mathbf{t}$ has an upward component. Hence, the angle θ is positive to the "left" of $\mathbf{k}$ in the plane of the profile curve.

There is also a unit normal $\mathbf{n}(s, v)$ to the surface $\mathbf{x}(s, v)$ determined as follows: the unit vector $\mathbf{t}$ and the vector $\mathbf{x}_v = r(s)\mathbf{e}_2$ give a basis for the tangent plane to $\mathbf{x}(s, v)$, so

$$\mathbf{n}(s, v) = \frac{\mathbf{t} \times \mathbf{x}_v}{|\mathbf{t} \times \mathbf{x}_v|} = -\cos\theta(s)\,\mathbf{e}_1(v) - \sin\theta(s)\,\mathbf{k}$$

is the desired unit normal. For a surface of revolution parameterized in the form $(h(s)\cos v, h(s)\sin v, g(s))$ and the unit normal specified above, we know (see Oprea 2007, Sect. 3.3.3) that the principal curvatures are given by

$$k_\mu = \frac{g''h' - g'h''}{(g'^2 + h'^2)^{3/2}}, \qquad k_\pi = \frac{g'}{h(g'^2 + h'^2)^{1/2}}. \tag{4.3}$$

The subscript μ denotes that k_μ is the curvature of the meridian (given by the intersection of the plane determined by $\mathbf{t}$ and $\mathbf{n}$ at any point and $\mathbf{x}$). The subscript π denotes that k_π is the curvature given by the intersection of the plane determined by $\mathbf{x}_v$ and $\mathbf{n}$ at any point and $\mathbf{x}$. For our surface $\mathbf{x}(s, v)$, we have $g = z$ and $h = r$, so we obtain

$$\begin{aligned}
k_\mu &= \frac{z''r' - z'r''}{1} = \left(\frac{-r'r''}{z'}\right)r' - z'r'' \\
&= \frac{-r''}{z'} = \frac{\cos\theta\,\theta'}{\cos\theta} = \theta',
\end{aligned}$$

where we have used $r'(s)^2 + z'(s)^2 = 1$ and $r'r'' + z'z'' = 0$ (by differentiating the first equation). We also obtain

$$k_\pi = \frac{z'}{r} = \frac{\cos\theta}{r}.$$

These principal curvatures will help us later on to understand the crucial interactions between tensions and geometry.

4.2.2 Tensions

Curved membranes do not necessarily change geometry to resist smoothly distributed loads. Furthermore, because shapes are determined at equilibrium, stresses may be found by solving differential equations (with boundary conditions). For arguments justifying these conclusions, see Irvine (1981, Sect. 5.1). So, our goal is to find the determining differential equations.

There are three possible tensions to consider: the *meridian stress* σ_m in the direction $\mathbf{t}$, the *circumferential* (or *hoop*) stress σ_c in the direction $\mathbf{e}_2$, and the shear stress. As argued in Irvine (1981), for membranes that are surfaces of revolution, shear stresses are zero due essentially to symmetry about an axis. These internal tensions are given in units of force per unit length. An inflated membrane has an external pressure $p(s)\overline{\mathbf{n}}(s, v) - w(s)\mathbf{k}$, where $\overline{\mathbf{n}} = -\mathbf{n}$ is the outward normal, the pressure $p(s)$ depends only on the meridian parameter s by symmetry about the z-axis, and $w(s)$ is a weight density associated with the membrane itself. Note that pressure normally pushes the membrane outward, while weight is directed downward, as usual. Consider a patch on the membrane (see Fig. 4.2) with parameter bounds $\mathring{s} \leq s$ and $\mathring{v} \leq v$. The patch is in equilibrium, so the total force acting on it is zero. Instead of writing things componentwise, we use vector notation and write

$$0 = \int_{\mathring{v}}^{v} \sigma_m(s)r(s)\mathbf{t}(s, u)\,du - \int_{\mathring{v}}^{v} \sigma_m(\mathring{s})r(\mathring{s})\mathbf{t}(\mathring{s}, u)\,du$$

$$+ \int_{\mathring{s}}^{s} \sigma_c(t)\mathbf{e}_2(v)\,dt - \int_{\mathring{s}}^{s} \sigma_c(t)\mathbf{e}_2(\mathring{v})\,dt$$

$$+ \int_{\mathring{s}}^{s} \int_{\mathring{v}}^{v} p(t)r(t)\overline{\mathbf{n}}(t, u)\,du\,dt - \int_{\mathring{s}}^{s} \int_{\mathring{v}}^{v} w(t)r(t)\mathbf{k}\,du\,dt.$$

Now, take $\partial/\partial s$ on both sides to obtain

$$0 = \int_{\mathring{v}}^{v} \frac{\partial}{\partial s}(\sigma_m(s)r(s)\mathbf{t}(s, u))\,du + \sigma_c(u)\mathbf{e}_2(v) - \sigma_c(u)\mathbf{e}_2(\mathring{v})$$

$$+ \int_{\mathring{v}}^{v} p(u)r(u)\overline{\mathbf{n}}(s, u)\,du - \int_{\mathring{v}}^{v} w(s)r(s)\mathbf{k}\,du.$$

Then, take $\partial/\partial v$ on both sides of this equation to obtain

Fig. 4.2 A patch on an axisymmetric membrane which is in equilibrium under various forces acting on it

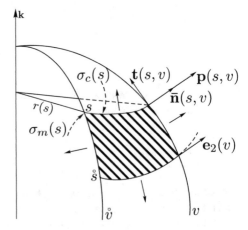

$$0 = \frac{\partial}{\partial s}(\sigma_m(s)r(s)\mathbf{t}(s, v)) - \sigma_c(s)\mathbf{e}_1(v) + p(s)r(s)\overline{\mathbf{n}}(s, v) - w(s)r(s)\mathbf{k}. \quad (4.4)$$

Now we can project onto $\mathbf{t}$ and $\overline{\mathbf{n}}$ by dotting with $\mathbf{t}$ and $\overline{\mathbf{n}}$, respectively. We use several facts: $\mathbf{t} \cdot \partial \mathbf{t}/\partial s = 0$, $\mathbf{e}_1 \cdot \mathbf{t} = -\sin\theta$ and

$$\frac{\partial}{\partial s}(\sigma_m r \mathbf{t}) \cdot \mathbf{t} + (\sigma_m r \mathbf{t}) \cdot \frac{\partial \mathbf{t}}{\partial s} = \frac{\partial}{\partial s}(\sigma_m r \mathbf{t} \cdot \mathbf{t})$$

$$= \frac{\partial}{\partial s}(\sigma_m r) + 2(\sigma_m r)\frac{\partial \mathbf{t}}{\partial s} \cdot \mathbf{t}$$

$$\frac{\partial}{\partial s}(\sigma_m r \mathbf{t}) \cdot \mathbf{t} = \frac{\partial}{\partial s}(\sigma_m r).$$

Therefore, we have the equations

$$0 = \frac{\partial}{\partial s}(\sigma_m r) + \sigma_c \sin\theta - wr\cos\theta$$

$$\frac{\partial}{\partial s}(\sigma_m r) = -\sigma_c \sin\theta + wr\cos\theta. \quad (4.5)$$

Dotting with $\overline{\mathbf{n}}$ gives us (using $\partial \mathbf{t}/\partial s = -\theta'\overline{\mathbf{n}}$)

$$0 = \frac{\partial}{\partial s}(\sigma_m r \mathbf{t}) \cdot \overline{\mathbf{n}} - \sigma_c \mathbf{e}_1 \cdot \overline{\mathbf{n}} + pr\overline{\mathbf{n}} \cdot \overline{\mathbf{n}} - wr\mathbf{k} \cdot \overline{\mathbf{n}}$$

$$0 = -\sigma_m r\theta' - \sigma_c \cos\theta + pr - wr\sin\theta \quad (4.6)$$

$$\sigma_m r\theta' = pr - \sigma_c \cos\theta - wr\sin\theta.$$

4.2.3 The Case $w = 0$

Let us consider the case in which the weight of the membrane is negligible, that is, $w \equiv 0$. Recall that $k_\mu = \theta'$ and $k_\pi = \cos\theta/r$. From (4.6), we get

$$\sigma_m \theta' + \frac{\sigma_c}{r}\cos\theta = p$$

$$\sigma_m k_\mu + \sigma_c k_\pi = p.$$

If we define the radii of curvature by $r_\mu = 1/k_\mu$ and $r_\pi = 1/k_\pi$, then we have a version of the *Laplace–Young equation* (see, for instance, Oprea 2000)

$$\frac{\sigma_m}{r_\mu} + \frac{\sigma_c}{r_\pi} = p. \quad (4.7)$$

Remark 4.1 Of course, when $w \neq 0$, we then have

$$\frac{\sigma_m}{r_\mu} + \frac{\sigma_c}{r_\pi} = p - \frac{w}{r}\sin\theta. \tag{4.8}$$

Now, when $w = 0$, Eq. (4.5) becomes $\frac{\partial}{\partial s}(\sigma_m r) = -\sigma_c \sin\theta$. Since $r' = -\sin\theta$, a solution is given by

$$\sigma_m = \sigma_c = \sigma = \text{ constant.}$$

Put this in the Laplace–Young equation (4.7) to get

$$\sigma\left(\theta' + \frac{\cos\theta}{r}\right) = p$$

$$\frac{1}{2}(k_\mu + k_\pi) = \frac{1}{2}\frac{p}{\sigma}$$

$$H = \frac{1}{2}\frac{p}{\sigma},$$

where H is the *mean curvature* of the membrane. If the pressure p is constant, then H is constant as well, and the membrane is a *surface of Delaunay* (see Oprea 2000).

It is also easy to conclude that when p is still constant, but $\sigma_m \neq \sigma_c$, Eq. (4.7) defines the so-called *anisotropic Delaunay surfaces* (for greater detail on this subject, the reader is referred to Koiso and Palmer (2008)).

Finally, if $p = 0$ and $\sigma_m \neq \sigma_c$, one ends up with the quite interesting class of *linear Weingarten surfaces* (Mladenov and Oprea 2003; Lopez 2008).

Consider (4.5) again (when $w = 0$) and suppose that $\sigma_c = 0$ (this situation is known in the literature as the *natural shape of the ballon*, see Baginski and Winker (2004) and Baginski (2005)), $p = \alpha$, constant. Then, $\sigma_m r = \beta$ is a constant as well and $\sigma_m = \beta/r$. From Eq. (4.6), we get (using $r' = -\sin\theta$)

$$\frac{\beta}{r}r\theta' = \alpha r$$

$$\beta\theta' = \alpha r$$

$$\theta' = \frac{\alpha}{\beta}r$$

$$2r'\theta' = 2\frac{\alpha}{\beta}rr'$$

$$-2\sin\theta\theta' = 2\frac{\alpha}{\beta}rr'$$

$$2(\cos\theta)' = \frac{\alpha}{\beta}(r^2)'$$

$$2\cos\theta = \frac{\alpha}{\beta}r^2 + d.$$

From our assumptions that the profile curve rises from the OX-axis and goes to the OZ-axis orthogonally, we see that $\theta = \pi/2$ exactly when $r = 0$. Hence, $d = 0$. Therefore, we have

$$2\frac{\cos\theta}{r} = \frac{\alpha}{\beta}r = \theta', \qquad 2k_\pi = k_\mu. \tag{4.9}$$

This condition will be explored in the next section. It describes the Mylar balloon. We can thus state the following:

Theorem 4.1 *If a membrane with* $w = 0$, *constant pressure* p *and hoop stress* $\sigma_c = 0$ *is a surface of revolution*

$$\mathbf{x}(s, v) = (r(s)\cos v, r(s)\sin v, z(s)) = r(s)\mathbf{e}_1(v) + z(s)\mathbf{e}_3(v),$$

then $2k_\pi = k_\mu$, *that is, the membrane is a Mylar balloon (see Sect. 4.4).*

4.2.4 Shapes and the Corresponding Surfaces

In the period between 1960 and 1970, J. Smalley did extensive work on axisymmetric balloon shapes and implemented these models on a digital computer.

As most of Smalley's considerations were of numerical origin, it warrants our looking for those models possessing analytical solutions. Despite the fact that the system governing these shapes is highly nonlinear, we have been successful in finding a few exact solutions, which are presented below (see Popova et al. 2006).

These solutions have been found neglecting some parameters in the equilibrium equations.

Let us start with the case in which we can neglect the film weight contribution, i.e., we suppose that $w(s) \equiv 0$, and hence, in such a case, instead of Eqs. (4.5) and (4.6), we have the system

$$\frac{\partial(\sigma_m r)}{\partial s} = -\sigma_c \sin\theta \tag{4.10}$$

$$(\sigma_m r)\theta' = pr - \sigma_c \cos\theta. \tag{4.11}$$

In order to be coherent with the geometrical relation (4.2), the first equation in the system implies that the meridional and circumferential stresses are constant and of the same magnitude, i.e., $\sigma_m = \sigma_c = \sigma = $ constant, while (4.11) specifies the mean curvature of $\mathcal{S}$, namely,

$$H = \frac{p}{2\sigma}. \tag{4.12}$$

4.3 Delaunay Surfaces

If we can arrange that the hydrostatic pressure is also a constant, i.e., $p(u) = p_o = $ const, then we end up with a surface of constant mean curvature

$$H = \frac{p_o}{2\sigma} = \text{const.} \tag{4.13}$$

Delaunay (1841) has isolated this class of surfaces guided by a genuine geometrical argument: that they are all just traces of the foci of the non-degenerate conics when they roll along a straight line in a plane (*roulettes* in French).

In the Appendix to the same paper, Sturm characterizes Delaunay's surfaces variationally as those surfaces of revolution having a minimal lateral area at a fixed volume. That, in turn, revealed why these surfaces make their appearance as soap bubbles and liquid drops or cells under compression, and now as membrane shapes. The complete list of Delaunay's surfaces includes cylinders of radius R and mean curvature $H = 1/2R$, spheres of radius R and mean curvature $H = 1/R$, catenoids of mean curvature $H = 0$, and nodoids and unduloids of constant non-zero mean curvatures. Their profile curves are shown in Fig. 4.3.

4.3.1 Nodoids and Unduloids

In this section, we will derive the analytical description of the last two and most interesting cases from Delaunay's list. We start with the system formed by Eqs. (4.10) and (4.11) which ensures the geometrical relation

$$\cos\theta = \frac{\mathring{p}r}{2} + \frac{C}{r}, \tag{4.14}$$

where C is some integration constant. Combined with (4.2), this leads to the equation

$$r' = -\sin\theta = -\frac{1}{2r}\sqrt{-\mathring{p}^2 r^4 + 4(1 - \mathring{p}C)r^2 - 4C^2}, \tag{4.15}$$

in which the variables can be separated, i.e.,

$$\frac{2r\,dr}{\sqrt{-\mathring{p}^2 r^4 + 4(1 - \mathring{p}C)r^2 - 4C^2}} = -ds. \tag{4.16}$$

Cylinder	Sphere	Catenoid	Unduloid	Nodoid

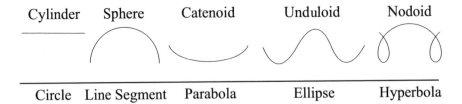

Circle	Line Segment	Parabola	Ellipse	Hyperbola

Fig. 4.3 The profile *curves* of Delaunay's surfaces obtained by rolling the conics listed below the horizontal *line*

The unpleasant sign on the right-hand side can be eliminated by going to a new variable, say $u = -s$, as we can measure the distance along the curve in only two ways. Additionally introducing a new variable $\xi = r^2$, we end up with the task for the evaluation of the elementary integral (on the left) written below

$$\int \frac{d\xi}{\sqrt{(c^2 - \xi)(\xi - a^2)}} = \int du = u + \phi, \tag{4.17}$$

where

$$a = \frac{1 - \sqrt{1 - 2\mathring{p}C}}{\mathring{p}}, \quad c = \frac{1 + \sqrt{1 - 2\mathring{p}C}}{\mathring{p}} \quad \text{and} \quad \phi \in \mathbb{R} \tag{4.18}$$

is some integration constant.

After some calculations, the result of integration can be written in the form

$$\xi(u) = r^2(u) = [(c^2 - a^2)\sin u + (c^2 + a^2)]/2, \tag{4.19}$$

in which the integration constant is omitted, as it is inessential for our further considerations.

In order to find the generating curve, we also have to solve the second equation in (4.2), which, in view of the above reads as

$$\frac{dz}{dr} = -\tan\theta = -\frac{\mathring{p}r^2 + 2C}{\sqrt{(c^2 - r^2)(r^2 - a^2)}}, \tag{4.20}$$

and, therefore

$$z(r) = -\int \frac{(\mathring{p}r^2 + 2C)dr}{\sqrt{(c^2 - r^2)(r^2 - a^2)}}. \tag{4.21}$$

The integral on the right-hand side can be uniformized by performing the change

$$r(t) = c\,\mathrm{dn}(t, k), \tag{4.22}$$

where $\mathrm{dn}(t, k)$ is one of the Jacobian elliptic functions of the argument t and the elliptic module k. Choosing k to be $\sqrt{c^2 - a^2}/c$, we get

$$z(t) = c\mathring{p} \int \mathrm{dn}^2(t, k)dt + \frac{2C}{c} \int dt, \tag{4.23}$$

and consequently

$$z(t) = c\mathring{p}E\left(\mathrm{am}(t, k), k\right) + \frac{2C}{c}F\left(\mathrm{am}(t, k), k\right). \tag{4.24}$$

Fig. 4.4 The open parts of the cylinder, sphere, catenoid, unduloid and nodoid shown here are drawn via the profile *curves* (4.32) and (4.33) or (4.37) and various combinations of the parameters λ and μ.

Finally, Eqs. (4.19) and (4.22) will be compatible if the natural parameter u and the uniformizing parameter t are related by the equation

$$\sin u = 1 - 2\text{sn}^2(t, k). \tag{4.25}$$

Let us also mention that another pair of formulas in place of (4.22) and (4.24) used for drawing the unduloid and nodoid in Fig. 4.4 has been derived following the variational approach in Mladenov and Oprea (2003a).

4.3.2 Intrinsic Equation of the Profile Curves of Delaunay Surfaces

For certain reasons (including typographical ones), it will be slightly useful to exchange the notation as follows:

$$H = \frac{p_o}{2\sigma} = \lambda = \text{constant}, \tag{4.26}$$

and $C = \mu$, so that Eq. (4.14) takes the form

$$\cos \theta = \lambda r + \frac{\mu}{r}. \tag{4.27}$$

The latter can be recognized as the Gauss map of the Delaunay surfaces (see Eells 1987). Without any loss of generality, we can assume that the constant λ is a strictly positive number relying either on physical experiments with membranes and balloons or taking into account the mathematical fact that $r \equiv r(s)$ is always positive and that we can measure $\theta \equiv \theta(s)$ only in two ways, clockwise or counterclockwise. The case in which $\lambda \equiv 0$ will be treated separately below.

Differentiating (4.27) consecutively with respect to s, we get

$$\theta' = \lambda - \frac{\mu}{r^2} \tag{4.28}$$

and

$$\theta'' = -\frac{2\mu}{r^3} \sin\theta. \qquad (4.29)$$

Taking into account that $\theta'(s)$ coincides with the curvature $\kappa \equiv \kappa(s) = \kappa_\mu(s)$ of the profile curve of the surface in the XOZ plane, Eq. (4.29) can be rewritten into the form

$$\kappa' = -2(\lambda - \kappa)\sqrt{\frac{\lambda - \kappa}{\mu} - (2\lambda - \kappa)^2}, \qquad (4.30)$$

which is just the intrinsic equation (Mladenov et al. 2008) of the meridional curve we have sought. As before, the minus sign in front of (4.30) again suggests changing the independent variable, i.e., $s = -u$. Respectively, the solution to Eq. (4.30) is

$$\kappa(u) = \lambda \frac{1 - 4\lambda\mu + \sqrt{1 - 4\lambda\mu}\sin(2\lambda u)}{1 - 2\lambda\mu + \sqrt{1 - 4\lambda\mu}\sin(2\lambda u)}, \qquad -\infty \le \mu \le \frac{1}{4\lambda}, \qquad (4.31)$$

which further implies via (4.28) that

$$r(u) = \frac{\sqrt{1 - 2\lambda\mu + \sqrt{1 - 4\lambda\mu}\sin(2\lambda u)}}{\lambda\sqrt{2}}. \qquad (4.32)$$

Integrating the second equation in (4.2) (in conjunction with (4.27) and (4.32)), we immediately obtain

$$z(u) = \frac{\mu}{m(\lambda, \mu)} F\left(\lambda u - \frac{\pi}{4}, k\right) + \frac{m(\lambda, \mu)}{\lambda} E\left(\lambda u - \frac{\pi}{4}, k\right), \qquad (4.33)$$

where the numerical factors are given by the expressions

$$m(\lambda, \mu) = \frac{\sqrt{1 - 2\lambda\mu + \sqrt{1 - 4\lambda\mu}}}{\sqrt{2}}, \qquad k = \sqrt{\frac{2\sqrt{1 - 4\lambda\mu}}{1 - 2\lambda\mu + \sqrt{1 - 4\lambda\mu}}}, \qquad (4.34)$$

while $F(\varphi, k)$ and $E(\varphi, k)$ denote the so-called incomplete elliptic integrals of, respectively, the first and second kinds, which are functions of their argument φ, and the parameter k is known as an elliptic modulus.

Let us also consider the case in which the differential hydrostatic pressure across the membrane vanishes, i.e., $\lambda \equiv 0$. In that case, the intrinsic equation (4.30) reduces to

$$\tilde{\kappa}' = 2\tilde{\kappa}\sqrt{-\frac{\tilde{\kappa}}{\mu} - \tilde{\kappa}^2}, \qquad (4.35)$$

and its solution is

$$\tilde{\kappa}(u) = -\frac{\mu}{u^2 + \mu^2}. \qquad (4.36)$$

This time, (4.28) and (4.27) produce the parameterization of the catenoid

$$\tilde{r}(u) = \sqrt{u^2 + \mu^2}, \quad \tilde{z}(u) = \mu\text{Ln}\left(u + \sqrt{u^2 + \mu^2}\right) \tag{4.37}$$

drawn in Fig. 4.4.

4.3.3 Some Useful Formulas

Having the explicit form of the parameterization (4.32) and (4.33) or (4.37), one can easily find any other geometrical characteristic of the surface $\mathcal{S}$. It is a general theorem in the classical differential geometry that for such a purpose, one needs to know only the first and the second fundamental forms of the surface under consideration. Actually, we have already derived the corresponding formulas for E, F, G, L and M, and further direct computations (in the $\lambda \neq 0$ case) produce

$$
\begin{aligned}
N &= \frac{1 + \sqrt{1 - 4\lambda\mu}\sin(2\lambda u)}{2\lambda} \\
\kappa_\pi(s) &= \lambda\frac{1 + \sqrt{1 - 4\lambda\mu}\sin(2\lambda u)}{1 - 2\lambda\mu + \sqrt{1 - 4\lambda\mu}\sin(2\lambda u)} .
\end{aligned}
\tag{4.38}
$$

The formulas mentioned above allows for an easy check that by taking $\mu = \dfrac{1}{4\lambda}$, one does indeed gets a cylinder, and that $\mu = 0$ corresponds to a sphere. The intermediate cases in which $0 < \mu < \dfrac{1}{4\lambda}$ generate unduloids, and those with $\mu < 0$ lead to nodoids.

If one is interested in the solution of the *inverse problem*, i.e., how to find the corresponding parameters λ, μ if the maximal $r_{\max}$ and minimal $r_{\min}$ distances from the symmetry axis are given, one easily ends up with the conclusion that in the case of the unduloid, these are

$$
\begin{aligned}
\lambda &= \frac{1}{r_{\max} + r_{\min}}, & \mu &= \frac{r_{\max} r_{\min}}{r_{\max} + r_{\min}} \\
r_{\max} &= \frac{\sqrt{1 - 2\lambda\mu} + \sqrt{1 - 4\lambda\mu}}{\lambda\sqrt{2}}, & r_{\min} &= \frac{\sqrt{1 - 2\lambda\mu} - \sqrt{1 - 4\lambda\mu}}{\lambda\sqrt{2}},
\end{aligned}
\tag{4.39}
$$

and the respective nodoid with the same geometrical data can be built with

$$\lambda = \frac{1}{r_{\max} - r_{\min}}, \quad \mu = -\frac{r_{\max} r_{\min}}{r_{\max} - r_{\min}} . \tag{4.40}$$

The parameters for the cylinders and spheres are recovered directly via (4.40), taking into account that their geometry is specified, respectively, by $r_{\max} = r_{\min}$ in the first case and $r_{\max} \in \mathbb{R}^+$, $r_{\min} = 0$ in the second.

All formulas above are also indispensable in such problems as finding the length, lateral surface area or the volume inside arbitrarily chosen segments of Delaunay surfaces (for more details, see Grason and Santangelo (2006) and Hadzhilazova et al. (2007)).

For the catenoids ($\lambda = 0$), one has, respectively, $E = 1$, $F = 0$, $G = u^2 + \mu^2$ for the first, and $L = -\dfrac{\mu}{u^2 + \mu^2}$, $M = 0$, $N = \mu$ for the second fundamental form, while by $\mu = \pm r_{\min}$, one recovers the explicit parameterization (4.37) of the surface.

4.3.4 Delaunay Construction

It was already mentioned that Delaunay starts by finding the evolute $\tilde{C}$ of the sought profile curve C. In what follows, we will keep to his notation as much as possible.

If ρ is the radius of the curvature of C and $\tilde{s}$ is the natural parameter on its evolute $\tilde{C}$, by its very definition, one has

$$\rho = \mathring{c} - \tilde{s}, \tag{4.41}$$

where $\mathring{c}$ is an arbitrarily real parameter. Let $\mathfrak{n}$ denotes the part of the tangent $\tilde{\mathbf{T}}$ to $\tilde{C}$ between M and its intersection with the symmetry axis OX. The condition that the surface S obtained by revolving C about OX has a constant mean curvature $\dfrac{1}{2a}$ in which a is another real parameter can be written as

$$\frac{1}{\rho} + \frac{1}{\mathfrak{n}} = \frac{1}{a}. \tag{4.42}$$

By expecting Fig. 4.5, one also easily finds that

Fig. 4.5 The *curve C* and it's evolute $\tilde{C}$

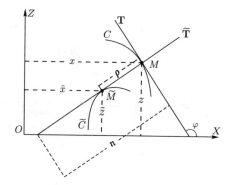

$$\mathbf{n} = \tilde{z}\frac{d\tilde{s}}{d\tilde{z}} + \overset{\circ}{c} - \tilde{s},$$ (4.43)

and therefore

$$\frac{1}{\overset{\circ}{c} - \tilde{s}} + \frac{1}{\tilde{z}\frac{d\tilde{s}}{d\tilde{z}} + \overset{\circ}{c} - \tilde{s}} = \frac{1}{a}.$$ (4.44)

Integrating the last equation, one gets

$$\tilde{z}^2 = \alpha(\overset{\circ}{c} - \tilde{s})(2a - \overset{\circ}{c} + \tilde{s}),$$ (4.45)

where α is the integration constant. This equation can be solved for $\tilde{s}$, and after that, the result differentiated with respect to $\tilde{z}$ in order to obtain

$$\frac{d\tilde{z}}{d\tilde{s}} = -\frac{\alpha\sqrt{a^2 - \tilde{z}^2/\alpha}}{\tilde{z}},$$ (4.46)

along with

$$\frac{d\tilde{x}}{d\tilde{s}} = \sqrt{1 - (\frac{d\tilde{z}}{d\tilde{s}})^2} = \frac{\sqrt{(1 + \alpha)\tilde{z}^2 - a^2\alpha^2}}{\tilde{z}}.$$ (4.47)

An inspection of (4.46) and (4.47) leads to the conclusion that in the above expressions, the constant α can take all positive values, and that in this case, $\tilde{z}$ will vary in the interval $[\frac{a\alpha}{\sqrt{1 + \alpha}}, a\sqrt{\alpha}]$. If α is negative, it can take values between -1 and 0, while $|\tilde{z}|$ can take any value greater than $-\frac{a\alpha}{\sqrt{1 + \alpha}}$. This means that in the last case, the evolute will have infinite branches. The two alternatives just described will be considered below separately. They are described in detail in Hadzhilazova and Mladenov (2009).

4.3.5 Nodary

Assuming that $0 < \alpha \leq \infty$, and introducing

$$m^2 = \frac{a^2\alpha^2}{1 + \alpha}, \qquad n^2 = a^2\alpha, \qquad n^2 > m^2.$$ (4.48)

Equations (4.46) and (4.47) can be combined into the form

$$\frac{d\tilde{x}}{d\tilde{z}} = -\frac{\sqrt{(1 + \alpha)\tilde{z}^2 - a^2\alpha^2}}{\alpha\sqrt{a^2 - \tilde{z}^2/\alpha}} = -\sqrt{\frac{1 + \alpha}{\alpha}}\frac{\sqrt{\tilde{z}^2 - m^2}}{\sqrt{n^2 - \tilde{z}^2}}.$$ (4.49)

The last expression suggests that it can be uniformized via

$$\tilde{z} = \frac{m}{\mathrm{dn}(u,k)}, \qquad m = \frac{a\alpha}{\sqrt{1+\alpha}} = a\alpha k, \qquad k = \frac{1}{\sqrt{1+\alpha}}, \qquad (4.50)$$

where $\mathrm{dn}(u,k)$ is one of the three Jacobian elliptic functions, u is its argument and the parameter k is known as an elliptic modulus. Using relations (4.50) and Eq. (4.49), we have

$$\tilde{x} = m(u - E(\mathrm{am}(u,k),k)), \qquad (4.51)$$

where $\mathrm{am}(u,k)$ is Jacobi's amplitude function, $E(\psi,k)$ denotes the so-called incomplete elliptic integral of the second kind, and the integration constant is omitted. Taken together, (4.50) and (4.51) provide the explicit parameterization of the evolute $\tilde{C}$. Its involute, i.e., the profile curve C of the Delaunay surface of constant mean curvature $\frac{1}{2a}$, can be found relying on direct geometrical relations (or consulting such textbooks on classical differential geometry as Gray (1998) or Oprea (2000))

$$x = \tilde{x} + \rho \frac{d\tilde{x}}{d\tilde{s}}, \qquad z = \tilde{z} + \rho \frac{d\tilde{z}}{d\tilde{s}}. \qquad (4.52)$$

By (4.41), (4.45)–(4.47) and (4.50), one easily finds

$$\rho = a - \sqrt{a^2 - \frac{\tilde{z}^2}{\alpha}} = a\left(1 - k\frac{\mathrm{cn}(u,k)}{\mathrm{dn}(u,k)}\right) \qquad (4.53)$$

$$\frac{d\tilde{x}}{d\tilde{s}} = \mathrm{sn}(u,k), \qquad \frac{d\tilde{z}}{d\tilde{s}} = -\mathrm{cn}(u,k),$$

which, taken together, give us the parameterization of the nodary

$$x[u] = m(u - E(\mathrm{am}(u,k),k)) + a\left(1 - k\frac{\mathrm{cn}(u,k)}{\mathrm{dn}(u,k)}\right)\mathrm{sn}(u,k)$$

$$z[u] = \frac{m}{\mathrm{dn}(u,k)} - a\left(1 - k\frac{\mathrm{cn}(u,k)}{\mathrm{dn}(u,k)}\right)\mathrm{cn}(u,k). \qquad (4.54)$$

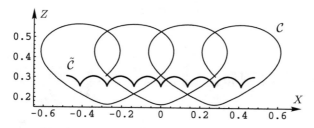

Fig. 4.6 The evolute $\tilde{C}$ of the *nodary* C generated with $\alpha = 2.333$ and $a = 0.2$ by (4.50), (4.51) and (4.54)

Both, the *nodary* C and its evolute $\tilde{C}$ are depicted in Fig. 4.6 for a concrete values of the parameters α and a.

4.3.6   Undulary

Following the plan announced above, we will consider in this section the case in which α belongs to the interval $-1 < \alpha < 0$. Because the treatment will be quite similar to that of a nodary the details will simply be outlined but in order to distinguish the cases, we will use the bars in the respective formulas that will remind us that we are dealing with negative α. In this setting, we will have

$$\frac{\mathrm{d}\bar{\tilde{x}}}{\mathrm{d}\bar{s}} = \frac{\sqrt{(1+\alpha)\bar{\tilde{z}}^2 - a^2\alpha^2}}{\bar{\tilde{z}}}, \qquad \frac{\mathrm{d}\bar{\tilde{z}}}{\mathrm{d}\bar{s}} = \frac{\sqrt{-\alpha}\sqrt{\bar{\tilde{z}}^2 - a^2\alpha}}{\bar{\tilde{z}}}, \qquad (4.55)$$

and, respectively,

$$\frac{\mathrm{d}\bar{\tilde{x}}}{\mathrm{d}\bar{\tilde{z}}} = \frac{1}{\sqrt{-\alpha}}\sqrt{\frac{(1+\alpha)\bar{\tilde{z}}^2 - a^2\alpha^2}{\bar{\tilde{z}}^2 - a^2\alpha}} = \sqrt{\frac{1+\alpha}{-\alpha}}\sqrt{\frac{\bar{\tilde{z}}^2 - \bar{m}^2}{\bar{\tilde{z}}^2 + \bar{n}^2}}, \qquad (4.56)$$

where

$$\bar{m}^2 = \frac{a^2\alpha^2}{1+\alpha}, \qquad \bar{n}^2 = -a^2\alpha, \qquad \bar{m}^2 + \bar{n}^2 = -\frac{a^2\alpha}{1+\alpha}. \qquad (4.57)$$

This time, the uniformization can be accomplished by dn and cn, i.e.,

$$\bar{\tilde{z}} = \sqrt{\bar{m}^2 + \bar{n}^2}\,\frac{\mathrm{dn}(u, \bar{k})}{\mathrm{cn}(u, \bar{k})} = \sqrt{\frac{-\alpha}{1+\alpha}}\,\frac{\mathrm{dn}(u, \bar{k})}{\mathrm{cn}(u, \bar{k})} \qquad (4.58)$$

$$\bar{k}^2 = \frac{\bar{n}^2}{\bar{m}^2 + \bar{n}^2} = 1 + \alpha.$$

Doing this, we obtain

$$\mathrm{d}\bar{\tilde{x}} = -a\frac{\mathrm{cn}^2(u, \bar{k})}{\mathrm{sn}^2(u, \bar{k})}\,\mathrm{d}u, \qquad (4.59)$$

and consequently

$$\bar{\tilde{x}} = a\left(E(\mathrm{am}(u, \bar{k}), \bar{k}) + \frac{\mathrm{cn}(u, \bar{k})\mathrm{dn}(u, \bar{k})}{\mathrm{sn}(u, \bar{k})} \right), \qquad (4.60)$$

in which case, as before, the integration constant is omitted. Further on, it is also easy to find that

$$\bar{\rho} = a(1 - \frac{1}{\bar{k}}\frac{1}{sn(u, \bar{k})}), \qquad \frac{d\tilde{x}}{d\bar{s}} = \bar{k}\frac{cn(u, \bar{k})}{dn(u, \bar{k})}, \qquad \frac{d\tilde{z}}{d\bar{s}} = \frac{\sqrt{-\alpha}}{dn(u, \bar{k})}, \qquad (4.61)$$

which immediately gives us

$$\bar{x}[u] = \tilde{x} + \bar{\rho}\frac{d\tilde{x}}{d\bar{s}} = a(\bar{k} - \frac{1}{sn(u, \bar{k})})\frac{cn(u, \bar{k})}{dn(u, \bar{k})}$$

$$\bar{z}[u] = \tilde{z} + \bar{\rho}\frac{d\tilde{z}}{d\bar{s}} = \frac{a\sqrt{-\alpha}}{\bar{k}}\left(\frac{dn(u, \bar{k})}{sn(u, \bar{k})} + a(\bar{k} - \frac{1}{sn(u, \bar{k})})\frac{1}{dn(u, \bar{k})}\right). \qquad (4.62)$$

4.3.6.1 Remarks

The generating curves of the nodoids and the unduloids referred to by Eells (1987) as, respectively, *nodary* and *undulary*, and their evolutes, have been found following the original Delaunay construction. These curves are periodic along the symmetry axis and have one local minimum and one local maximum in each period; they do not depend on the chosen point on the evolutes, as the constant $\mathring{c}$ disappears from all formulae (see Fig. 4.7).

The parameterizations (up to integration) found by Delaunay himself (cf Delaunay 1841) are given below for a comparison with those derived here and elsewhere (Mladenov 2002a and Hadzhilazova et al. 2007), i.e.,

$$\tilde{x} = -\frac{a\alpha\tan\varphi}{\sqrt{1 + \alpha - \sin^2\varphi}} + \int_0^\varphi \frac{a\alpha d\varphi}{\cos^2\varphi\sqrt{1 + \alpha - \sin^2\varphi}} \qquad (4.63)$$

$$\tilde{z} = \frac{a\alpha}{\sqrt{1 + \alpha - \sin^2\varphi}}, \qquad \varphi \in \mathbb{R}$$

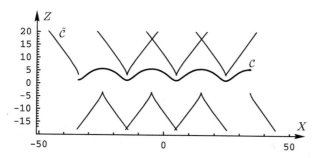

Fig. 4.7 The evolute $\tilde{C}$ of the *undulary* C generated with $\alpha = 3.5$ and $a = -0.6$ by (4.58), (4.60) and (4.62)

parameterize the evolutes, and

$$x[\varphi] = a \sin \varphi - a \tan \varphi \sqrt{1 + \alpha - \sin^2 \varphi} + \int_0^\varphi \frac{a\alpha d\varphi}{\cos^2 \varphi \sqrt{1 + \alpha - \sin^2 \varphi}} \qquad (4.64)$$

$$z[\varphi] = -a \cos \varphi + a \sqrt{1 + \alpha - \sin^2 \varphi}$$

do the same for the nodary and undulary. The parameters α and a have the same meaning as specified before.

Finally, it can easily be realized that the integrals which appear in Delaunay formulas exist only at restricted intervals in which the evolute can be found.

4.4 Mylar Balloon and Elastic Curves

This is another case in which the system of Eqs. (4.5) and (4.6) can be solved up to the very end. Assuming that $w(s) = \sigma_c = 0$ and that the hydrostatic pressure $p(s) = p_o$ is a constant (and non-zero), this system reduces to the equations

$$(\sigma_m(s)r(s)) \frac{d\theta(s)}{ds} = p_o r(s) \qquad (4.65)$$

$$\frac{d(\sigma_m(s)r(s))}{ds} = 0. \qquad (4.66)$$

The second equation above tells us that $\sigma_m(s)r(s)$ is a constant quantity, and therefore we can introduce the meridional stress resultant $\mathring{\sigma}$ on the equator of the balloon, i.e., the points for which $r(s) = a$, $z(s) = 0$, and rewrite the above integral in the form

$$\sigma_m(s) = \frac{a\mathring{\sigma}}{r(s)}. \qquad (4.67)$$

This also allows us to rewrite the first equation (4.65) as

$$\frac{d\theta(s)}{ds} = \mathring{p}r(s), \qquad \mathring{p} = \frac{p_o}{a\mathring{\sigma}} > 0. \qquad (4.68)$$

If we combine this equation with (4.2), we get the following geometrical relation:

$$r^2(s) = \frac{2}{\mathring{p}} \cos \theta(s). \qquad (4.69)$$

This last relation, as we shall see, uniquely characterizes the surface in question. Let us start by solving (4.69) for $r(s)$. After that, we replace the result in (4.68), and in this way obtain a differential equation with separated variables

$$\frac{d\theta}{\sqrt{\cos\theta}} = \sqrt{2\mathring{p}}\,ds.$$ (4.70)

Next, we introduce $\eta = \sin\theta$, which transforms the left-hand side into

$$\frac{d\eta}{\sqrt{\eta(1-\eta^2)}},$$

and this suggests a new change $\eta = \zeta^2$ of the independent variable η, which gives us

$$\frac{d\eta}{\sqrt{\eta(1-\eta^2)}} = \frac{2d\zeta}{\sqrt{1-\zeta^4}} = \frac{2d\zeta}{\sqrt{(1+\zeta^2)(1-\zeta^2)}}.$$

As a result of all these changes, Eq. (4.70) reduces to the form

$$\frac{2d\zeta}{\sqrt{(1+\zeta^2)(1-\zeta^2)}} = \sqrt{2\mathring{p}}\,ds.$$ (4.71)

However, this is a standard elliptic integral (see Jahnke et al. 1960), which can be inverted directly in Jacobi's eliptic functions cn, and we have

$$\zeta(s) = -\mathrm{cn}\left(\sqrt{\mathring{p}}s, \frac{1}{\sqrt{2}}\right).$$

As a consequence, we also have

$$\sin\theta(s) = \frac{\mathring{p}}{2}r^2(s) = \zeta^2(s) = \mathrm{cn}^2\left(\sqrt{\mathring{p}}s, \frac{1}{\sqrt{2}}\right),$$ (4.72)

and therefore

$$r(s) = \sqrt{\frac{2}{\mathring{p}}}\,\mathrm{cn}\left(\sqrt{\mathring{p}}s, \frac{1}{\sqrt{2}}\right).$$ (4.73)

In order to find $z(s)$, we make use of (4.20) and (4.72), which lead to

$$\frac{dz(s)}{ds} = -\mathrm{cn}^2\left(\sqrt{\mathring{p}}s, \frac{1}{\sqrt{2}}\right).$$ (4.74)

Details about the integration of the above equation can be found in Hadzhilazova and Mladenov (2006), and the result is

$$z(s) = -\frac{2}{\sqrt{\mathring{p}}}\left[E\left(\mathrm{am}\left(\sqrt{\mathring{p}}s, \frac{1}{\sqrt{2}}\right), \frac{1}{\sqrt{2}}\right) - \frac{1}{2}F\left(\mathrm{am}\left(\sqrt{\mathring{p}}s, \frac{1}{\sqrt{2}}\right), \frac{1}{\sqrt{2}}\right)\right].$$ (4.75)

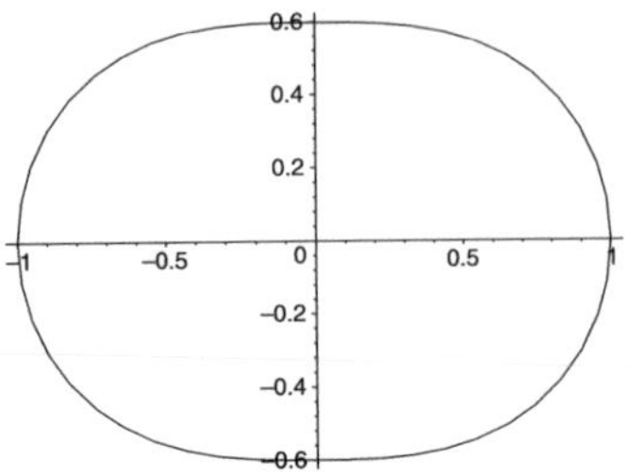

Fig. 4.8 The profile of the mylar balloon in the *XOZ* plane

Fig. 4.9 An open part of the mylar balloon

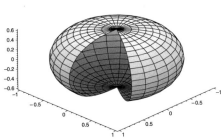

If we compare the obtained parameterization of the profile curve $(r(s), z(s))$ provided by (4.73) and (4.75) with that in Mladenov and Oprea (2003), we can easily conclude that we are dealing here with the Mylar balloon. The profile curve and the surface generated by them are shown in Figs. 4.8 and 4.9.

For commercial purposes, the just-mentioned Mylar balloon is fabricated from two circular disks of mylar, sewing them along their boundaries and then inflating team. Surprisingly enough, these balloons are not spherical, as one naively might expect from the well-known fact that the sphere possesses the maximal volume for a given surface area. An experimental fact like this suggests a mathematical problem regarding the exact shape of the balloon when it is fully inflated.

This problem was first spelled out by Paulsen (1994) in a variational setting, while here we have, in fact, provided its non-variational characterization. One should also mention the remarkable scale invariance (i.e., independence of the actual size) of the thickness to diameter ratio of the inflated balloon, which turns out to be a good approximation equal to 0.599 (see Mladenov 2002a). Another important fact about this surface is the very simple expression for its area given by the formula $\mathcal{A} = \pi^2 a^2$, where a is the radius of the inflated balloon. In some sense, all these nice properties are due to the remarkable property which uniquely specifies the mylar balloon as the only surface of revolution for which the principal curvatures k_μ and k_π obey the equation

$$k_\mu = 2k_\pi. \tag{4.76}$$

As has been noticed by Gibbons (2006), this (Weingarten) property can be derived within the membrane approach as well, by rewriting (4.69) in the form

$$- \mathring{p}r(s) = -2\frac{\cos\theta(s)}{r(s)}. \tag{4.77}$$

Taking into account (4.68) along with the definitions of the principal curvatures given in (4.3) amounts directly to the equality (4.76).

Detailed differential-geometric proofs of the other facts mentioned above and many additional comments can be found in the already cited papers by Hadzhilazova and Mladenov (2006) and Popova et al. (2006).

4.4.1 Bending Energy

When one considers the elastic properties of the materials, the main question is: What is the energy needed to bend a rod in the plane? According to the theory of elasticity (see Love 1944), the energy is proportional to the integral of the squared curvature along its length

$$U = \int_0^L \frac{EI}{\mathcal{R}^2}ds, \tag{4.78}$$

where U is the bending energy, E is the Young's module of elasticity, I is the moment about the neutral axis, $\mathcal{R}$ is the curvature radius of the neutral axis, and L is its length.

Its study was initiated in 1691 by James Bernoulli who had tried to develop this model by using the available mechanics, geometry and variational calculus, an approach that was continued in Euler's book Euler (1744), which lays down the basis for modern variational calculus.

There, he examines the problem of finding the shape of an elastic ribbon with a fixed length L, which connects two points in the plane at which it has fixed tangents (see Fig. 4.10). We assume that the origin O of the coordinate system XOZ coincides with one of the points and that the other has coordinates (x, z).

Now our task is to find the minimum of the functional

$$J_0 = \int_0^L \kappa^2(s) \tag{4.79}$$

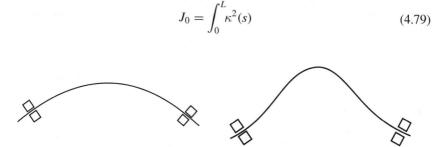

Fig. 4.10 Some possible positions of the ribbon when it is free, on the *left*, and with a fixed length, on the *right*

under the condition of a fixed length

$$\int_0^L ds = L. \tag{4.80}$$

Bernoulli and Euler make use of Eq. (1.12) and J_0 consequently reads as

$$\int \frac{\ddot{z}^2(x)}{(1 + \dot{z}^2)^{5/2}} dx, \qquad \dot{z} = \frac{dz}{dx}, \qquad \ddot{z} = \frac{d^2z}{dx^2}. \tag{4.81}$$

Despite the fact that, in this case, the Euler–Lagrange equation is of the fourth order, Euler managed to integrate it three times and to find the first order equation

$$\frac{dz}{dx} = \frac{a^2 - c^2 + x^2}{\sqrt{(c^2 - x^2)(2a^2 - c^2 + x^2)}}, \qquad a, c \in \mathbb{R}. \tag{4.82}$$

From this point on, his considerations were qualitative (the elliptic functions had not been invented yet!) and the solutions to (4.82) were classified according to the different values of the real parameter

$$m = \frac{a^2}{2c^2}. \tag{4.83}$$

In the next section, we will expose the above result by following Euler's original method.

4.4.2 Original Formulation and Treatment of the Problem About Elastic Curves

In this section, we will consider the original Euler method, which aims for the integration of the equation that describes the elasticas. As mentioned before, it is a nonlinear fourth order ordinary differential equation.

Let us start by writing the arc length ds in the form

$$ds = (dx^2 + dz^2)^{1/2} = (dx^2 + \dot{z}^2 dx^2)^{1/2} = (1 + \dot{z}^2)^{1/2} dx, \tag{4.84}$$

and for the full functional J, we can respectively write

$$J = \int \frac{\ddot{z}^2 dx}{(1 + \dot{z}^2)^{5/2}} + \lambda \int (1 + \dot{z}^2)^{1/2} dx.$$

According to (1.138), the Euler–Lagrange equation is

$$\frac{d^2}{dx^2}\left(\frac{\partial F}{\partial \ddot{z}}\right) - \frac{d}{dx}\left(\frac{\partial F}{\partial \dot{z}}\right) + \frac{\partial F}{\partial z} = 0,$$

and in our case, it is

$$\frac{d^2}{dx^2}(2\ddot{z}(1+\dot{z}^2)^{-5/2}) + \frac{d}{dx}(5\ddot{z}^2\dot{z}(1+\dot{z}^2)^{-7/2} - \lambda\dot{z}(1+\dot{z}^2)^{-1/2}) = 0.$$

Integrating the last equation once, we have

$$\frac{d}{dx}(2\ddot{z}(1+\dot{z}^2)^{-5/2}) + 5\ddot{z}^2\dot{z}(1+\dot{z}^2)^{-7/2} - \lambda\dot{z}(1+\dot{z}^2)^{-1/2} = A,$$

where A is an integration constant. Performing the differentiation in the above formula, we have

$$2\dddot{z}(1+\dot{z}^2)^{-5/2} - 5\ddot{z}^2\dot{z}(1+\dot{z}^2)^{-7/2} + \lambda\dot{z}(1+\dot{z}^2)^{-1/2} = A. \qquad (4.85)$$

In this equation, we can rewrite the $2\dddot{z}$ that appears in the first term as

$$2\dddot{z} = 2\ddot{z}\frac{d\ddot{z}}{d\dot{z}} = \frac{d\ddot{z}^2}{d\dot{z}},$$

the second one as

$$\ddot{z}^2\frac{d}{d\dot{z}}(1+\dot{z}^2)^{-5/2},$$

and the third one as

$$-\lambda\frac{d}{d\dot{z}}(1+\dot{z}^2)^{1/2}.$$

Doing so, Eq. (4.85) transforms into the form

$$(1+\dot{z}^2)^{-5/2}\frac{d\ddot{z}^2}{d\dot{z}} + \ddot{z}^2\frac{d}{d\dot{z}}(1+\dot{z}^2)^{-5/2} - \lambda\frac{d}{d\dot{z}}(1+\dot{z}^2)^{1/2} = A.$$

Now, we can unify the first two terms in the last equation to obtain

$$\frac{d}{d\dot{z}}(\ddot{z}^2(1+\dot{z}^2)^{-5/2}) - \frac{d}{d\dot{z}}(\lambda(1+\dot{z}^2)^{1/2}) = A. \qquad (4.86)$$

After a direct integration of (4.86), we get

$$\ddot{z}^2(1+\dot{z}^2)^{-5/2} = \lambda(1+\dot{z}^2)^{1/2} + A\dot{z} + B, \qquad (4.87)$$

where B is the new integration constant.

Solving the last equation for $\ddot{z}$ yields

$$\ddot{z} = \frac{d\dot{z}}{dx} = (1 + \dot{z}^2)^{5/4} \left(\lambda(1 + \dot{z}^2)^{1/2} + A\dot{z} + B \right)^{1/2}. \tag{4.88}$$

A crucial moment is the observation (due to Euler) that

$$\frac{d}{d\dot{z}} \left(\frac{2(\lambda(1 + \dot{z}^2)^{1/2} + A\dot{z} + B)^{1/2}}{(1 + \dot{z}^2)^{1/4}} \right) = \frac{A - B\dot{z}}{(1 + \dot{z}^2)^{5/4} \left(\lambda(1 + \dot{z}^2)^{1/2} + A\dot{z} + B \right)^{1/2}},$$

and taking into account (4.88), we end up with

$$\frac{d}{d\dot{z}} \left(\frac{2 \left(\lambda(1 + \dot{z}^2)^{1/2} + A\dot{z} + B \right)^{1/2}}{(1 + \dot{z}^2)^{1/4}} \right) = (A - B\dot{z}) \frac{dx}{d\dot{z}}. \tag{4.89}$$

The integration of the above equation is immediate and produces

$$\frac{2(\lambda(1 + \dot{z}^2)^{1/2} + A\dot{z} + B)^{1/2}}{(1 + \dot{z}^2)^{1/4}} = Ax - Bz + C, \tag{4.90}$$

where C is the integration constant. By making a special choice for the constants B and C, i.e., $B = C = 0$, and solving the so-reduced Eq. (4.90) for $\dot{z}$, we obtain the equation

$$\dot{z}(x) = \frac{A^2 x^2 - 4\lambda}{\sqrt{16A^2 - \left(A^2 x^2 - 4\lambda \right)^2}}. \tag{4.91}$$

At this stage, it is convenient to introduce the real numbers a and c as new parameters via the relations

$$a^4 = \frac{16}{A^2} \quad \text{and} \quad c^2 - a^2 = \frac{4\lambda}{A^2}, \tag{4.92}$$

and in this way to convert (4.91) into

$$\frac{dz}{dx} = \frac{a^2 - c^2 + x^2}{\sqrt{(c^2 - x^2)(2a^2 - c^2 + x^2)}}$$

which is exactly the Eulerian equation (4.82).

Alternatively, recalling the formula for the curvature

$$\kappa(x) = \frac{\ddot{z}}{(1 + \dot{z}^2)^{3/2}},$$

we can transform Eq. (4.87) into the form

$$\kappa^2 = A\frac{\dot{z}}{\sqrt{1 + \dot{z}^2}} + \frac{B}{\sqrt{1 + \dot{z}^2}} + \lambda. \tag{4.93}$$

According to Fig. 1.1, one has $\dot{z} = \dfrac{dz}{dx} = \tan\psi$, where $\psi = \psi(x)$ is the angle between the tangent at this point and the positive direction of OX axis, and hence Eq. (4.93) becomes

$$\kappa^2 = A\sin\psi + B\cos\psi + \lambda.$$

Differentiating this equality with respect to the arclength s (which we will denote with a prime) gives us

$$2\kappa\kappa' = A\cos\psi\psi' - B\sin\psi\psi'. \tag{4.94}$$

By its very definition,

$$\kappa = \frac{d\psi}{ds} = \psi',$$

and therefore we have the equation

$$2\kappa' = A\cos\psi - B\sin\psi, \tag{4.95}$$

in which, for the left-hand side we can write

$$\kappa' = \frac{d\kappa}{ds} = \frac{d\kappa}{dx}\frac{dx}{ds} = \frac{1}{\sqrt{1 + \dot{z}^2}}\frac{d\kappa}{dx} = \frac{\dot{\kappa}}{\sqrt{1 + \dot{z}^2}}.$$

In Cartesian coordinates, (4.95) becomes

$$\frac{2\dot{\kappa}}{\sqrt{1 + \dot{z}^2}} = \frac{A}{\sqrt{1 + \dot{z}^2}} - \frac{B\dot{z}}{\sqrt{1 + \dot{z}^2}},$$

which is obviously equivalent to the equation

$$2\dot{\kappa} = A - B\dot{z}.$$

Integrating the above relation, we get the equation of the Eulerian elasticas

$$\kappa(x, z) = \frac{A}{2}x - \frac{B}{2}z + C, \qquad C = \text{const}, \tag{4.96}$$

which tells us that elasticas are curves the curvatures of which are linear functions of the Cartesian coordinates. If we take as the ordinate the line

$$Bx + Az + D = 0$$

and for abscissa the perpendicular one

$$Ax - Bz + 2C = 0,$$

then the expression for the curvature of the elasticas (4.96) transforms into

$$\kappa(x) = \alpha x, \qquad \alpha \in \mathbb{R}, \tag{4.97}$$

where we have used the same notation for the new quantities. In Sect. 6.4.2, we will generate the elasticas using (4.97) as a definition. In the next section, we present an alternative method of derivation of (4.96) that has its own merits.

Remark 4.2 One can check the compatibility of the approaches presented above by evaluating the curvature using (4.82), which gives (modulo some calculations) the expected formula

$$\kappa(x) = \frac{2}{a^2}x. \tag{4.98}$$

4.4.3 Parametric Representation of the Curvature of Elastica

According to the alternative formula (1.24) for the curvature of plane curves, Euler's elasticas problem reduces to the study of the functional

$$J = \int (\ddot{x}^2 + \ddot{z}^2 + \nu(s)(\dot{x}^2 + \dot{z}^2))ds.$$

The over dot here means a differentiation with respect to the natural parameter. Respectively, the Euler–Lagrange equations 1.139 in this case reduce to the system

$$\frac{d^2\ddot{x}}{ds^2} - \frac{d}{ds}(\nu\dot{x}) = 0, \qquad \frac{d^2\ddot{z}}{ds^2} - \frac{d}{ds}(\nu\dot{z}) = 0,$$

which can be rewritten in their equivalent form

$$\frac{d}{ds}(\dddot{x} - \nu\dot{x}) = 0, \qquad \frac{d}{ds}(\dddot{z} - \nu\dot{z}) = 0.$$

From the formulas above, we can immediately obtain the equations

$$\dddot{x} - \nu\dot{x} = A, \qquad \dddot{z} - \nu\dot{z} = B, \tag{4.99}$$

where A and B are some integration constants. Multiplying the first equation with $\dot{z}$, the second with $\dot{x}$, and subtracting the second result from the first one, we obtain

$$\dddot{x}\dot{z} - \dot{x}\dddot{z} = A\dot{z} - B\dot{x}.$$

Furthermore, integrating once again gives us

$$\ddot{x}\dot{z} - \dot{x}\ddot{z} = Az - Bx + C.$$

The left-hand side of the above equality can be recognized as the curvature (see Eq. 1.14) in terms of the natural parameter s.

4.4.4 Intrinsic Equation of the Elastica

As we know, this means finding the equation connecting the curvature κ and the natural parameter s of the curve. For that purpose, we will use the Frenet–Serret equations

$$\dot{\mathbf{x}}(s) = \mathbf{T}(s), \qquad \mathbf{N} = \mathbf{T}^{\perp}, \qquad \dot{\mathbf{T}} = \kappa\mathbf{N}, \qquad \dot{\mathbf{N}} = -\kappa\mathbf{T}. \tag{4.100}$$

By definition, $\mathcal{C}$ is an elastic curve when the functional of energy $\int \kappa^2(s)ds$ has a minimum for a fixed length $L = \int ds$. Let us consider its infinitesimal deformation defined by the formula

$$\tilde{\mathbf{x}}(s) = \mathbf{x}(s) + \varepsilon(s)\mathbf{N}, \tag{4.101}$$

where $\varepsilon(s)$ is an infinitesimal function of the parameter s. Then,

$$\dot{\tilde{\mathbf{x}}} = \dot{\mathbf{x}} + \dot{\varepsilon}\mathbf{N} + \varepsilon\dot{\mathbf{N}} = \mathbf{T} + \dot{\varepsilon}\mathbf{N} - \varepsilon\kappa\mathbf{T} = (1 - \varepsilon\kappa)(\mathbf{T} + \dot{\varepsilon}\mathbf{N}). \tag{4.102}$$

For the last equation, we use the fact that $\varepsilon.\dot{\varepsilon}$ is a negligible quantity. Following the same strategy, we have

$$d\tilde{s} = (d\tilde{\mathbf{x}}.d\tilde{\mathbf{x}})^{1/2} = (\dot{\tilde{\mathbf{x}}}.\dot{\tilde{\mathbf{x}}})^{1/2}ds$$
$$= (1 - \varepsilon\kappa)[(\mathbf{T} + \dot{\varepsilon}\mathbf{N}).(\mathbf{T} + \dot{\varepsilon}\mathbf{N})]^{1/2}ds = (1 - \varepsilon\kappa)ds,$$

and therefore

$$\tilde{\mathbf{T}} = \frac{d\tilde{\mathbf{x}}}{d\tilde{s}} = \frac{\dot{\tilde{\mathbf{x}}}ds}{d\tilde{s}} = \frac{(1 - \varepsilon\kappa)(\mathbf{T} + \dot{\varepsilon}\mathbf{N})ds}{(1 - \varepsilon\kappa)ds} = \mathbf{T} + \dot{\varepsilon}\mathbf{N}.$$

The normal vector $\tilde{\mathbf{N}}$ to the deformed curve $\tilde{\mathcal{C}}$ is defined by the condition that it is orthogonal to $\tilde{\mathbf{T}}$ and that the (infinitesimal) deformation is along the direction of $\mathbf{N}$, i.e.,

$$\tilde{\mathbf{N}} = \mathbf{N} + \lambda\mathbf{T},$$

where λ is a function of s. According to the above-said we have

$$\tilde{\mathbf{N}}.\tilde{\mathbf{T}} = (\mathbf{N} + \lambda\mathbf{T})(\mathbf{T} + \dot{\varepsilon}\mathbf{N}) = \lambda + \dot{\varepsilon} = 0,$$

and therefore

$$\tilde{\mathbf{N}} = \mathbf{N} - \dot{\varepsilon}\mathbf{T}.$$

Imposing the condition that the deformed curve has a fixed length, i.e.,

$$\int d\tilde{s} = \int (1 - \varepsilon\kappa)ds = \int ds,$$

leads to its analytical form, presented by the equation

$$\int \varepsilon\kappa ds = 0.$$

According to the Frenet–Serret formulas (4.100), another equation has to be satisfied as well, i.e.,

$$\frac{d\tilde{\mathbf{T}}}{d\tilde{s}} = \tilde{\kappa}\tilde{\mathbf{N}} = \tilde{\kappa}(\mathbf{N} - \dot{\varepsilon}\mathbf{T}).$$

The left-hand side can be rewritten as

$$\frac{d\tilde{\mathbf{T}}}{d\tilde{s}} = \frac{(\mathbf{T} + \dot{\varepsilon}\mathbf{N})' ds}{(1 - \varepsilon\kappa)ds} = (1 + \varepsilon\kappa)(\kappa\mathbf{N} + \ddot{\varepsilon}\mathbf{N} - \dot{\varepsilon}\kappa\mathbf{T}) = (\kappa + \ddot{\varepsilon} + \varepsilon\kappa^2)(\mathbf{N} - \dot{\varepsilon}\mathbf{T}),$$

which means that the curvature of the deformed curve is given by the formula

$$\tilde{\kappa} = \kappa + \varepsilon\kappa^2 + \ddot{\varepsilon}.$$

Subsequently, as $\tilde{\mathcal{C}}$ is an extremal, we have

$$\begin{aligned}
\int \tilde{\kappa}^2 d\tilde{s} + 2\sigma \int d\tilde{s} &= \int (\kappa + \varepsilon\kappa^2 + \ddot{\varepsilon})^2(1 - \varepsilon\kappa)ds + 2\sigma \int (1 - \varepsilon\kappa)ds \\
&= \int \kappa^2 ds + 2\sigma \int ds + \int (2\kappa\ddot{\varepsilon} + \varepsilon\kappa^3)ds - 2\sigma \int \varepsilon\kappa ds \\
&= \int \kappa^2 ds + 2\sigma \int ds + \int (2\ddot{\kappa} + \kappa^3 - 2\sigma\kappa)\varepsilon ds.
\end{aligned}$$

In order to obtain the last formula, we have integrated twice by parts $\kappa\ddot{\varepsilon}$. Because $\mathcal{C}$ is an elastic curve, the integral $\int \kappa^2 ds$ has a minimal value for all deformations which preserve its length $\int ds$, and therefore we have the equation

$$2\ddot{\kappa} + \kappa^3 - 2\sigma\kappa = 0, \tag{4.103}$$

where σ is a constant. A direct consequence of the above equation is that the curve $\mathcal{C}$ is a critical point for the elastica's functional $\int (k^2 + 2\sigma)\mathrm{d}s$. We are not going to discuss the question as to which critical points are minimum and which are not, but we will present another derivation of Eq. (4.103) later on. Before that, in the next section, we will apply the so-developed variational techniques to a problem which is quite interesting in both its geometrical and mechanical aspects.

4.4.5 A Hanging Chain

The potential energy of any infinitesimal element of a homogeneous freely hanging heavy chain is proportional to the length $\mathrm{d}s$ of this element and its height, i.e.,

$$\mathrm{d}U = z\mathrm{d}s.$$

For the entire segment between the points A and B, the potential energy is given by the integral

$$J = \int_A^B z\mathrm{d}s,$$

and it is assumed that the linear density of the chain mass and the gravitation constant are equal to one. The equilibrium state of the chain is defined as the stationary points of J, i.e., the points for which

$$\delta J = 0, \tag{4.104}$$

where δ is the infinitesimal variation of the functional J (Fig. 4.11).

Before finding them, we will rewrite the results derived in previous section in an equivalent form more useful in the case under consideration, which reads as follows:

$$\delta \mathbf{x} = \varepsilon(s)\mathbf{N}(s), \qquad \delta \mathrm{d}s = -\varepsilon\kappa\mathrm{d}s, \qquad \delta \mathbf{T} = \dot{\varepsilon}\mathbf{N}, \qquad \delta \mathbf{N} = -\dot{\varepsilon}\mathbf{T}.$$

Fig. 4.11 A homogeneous chain in a gravitational field

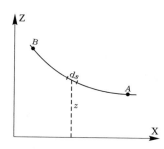

Concerning Eq. (4.104) above, we have

$$\delta J = \int \delta z ds + \int z \delta ds = \int \varepsilon(s) N_z ds - \int z \varepsilon \kappa ds$$
$$= \int \varepsilon(s)(N_z - z\kappa)ds = 0.$$

As $\varepsilon(s)$ is arbitrary function, the last equation is equivalent to

$$N_z - z\kappa = 0.$$

According to Fig. 1.1, $N_z = \cos \psi$, and after dividing by κ and differentiating with respect to s, we have

$$\frac{d}{ds}\left(\frac{\cos \psi}{\kappa} - z\right) = -\frac{\sin \psi}{\kappa}\dot{\psi} - \frac{\dot{\kappa}\cos \psi}{\kappa^2} - \sin \psi = 0.$$

Taking into account that $\dot{\psi} = \kappa$, we can rewrite this equation as

$$-2\tan \psi = \frac{\dot{\kappa}}{\kappa^2} = \frac{1}{2\kappa^2}\frac{d\kappa^2}{d\psi}, \qquad (4.105)$$

and integrating to obtain the relation

$$\kappa = c \cos^2 \psi, \qquad (4.106)$$

in which c is the integration constant. Combining (4.105) and (4.106) allows us to find the intrinsic equation of the curve in the form

$$\dot{\kappa}^2 + 4\kappa^4 - 4c^2\kappa^2 = 0.$$

The solution to this equation is

$$\kappa = \frac{c}{1 + c^2 s^2},$$

while for $\psi = \int \kappa ds$, we have

$$\psi = \arctan cs,$$

and, respectively,

$$\sin \psi = \frac{cs}{\sqrt{1 + c^2 s^2}}, \qquad \cos \psi = \frac{1}{\sqrt{1 + c^2 s^2}}.$$

Two additional integrations produce the coordinates of the points on the curve,

Fig. 4.12 The graphics of the chain generated via formula (4.107) for $c = 1.24$

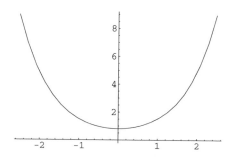

$$x(s) = \frac{\text{arc sinh}(cs)}{c}, \qquad z(s) = \frac{\sqrt{1 + c^2 s^2}}{c}.$$

If we eliminate the parameter s from these two equations, we have the standard formula in the textbooks

$$z(x) = \frac{\sqrt{1 + \sinh^2(cx)}}{c} = \frac{\cosh(cx)}{c}. \tag{4.107}$$

The plot of this function, i.e., the curve describing the form of the chain, is presented in Fig. 4.12.

4.4.6 Eulers Elasticas as One-Dimensional Membranes

As it was explained earlier, our problem is to find the extremum of the functional

$$J = \int_0^L \kappa^2(s)ds + \lambda_1 \int_0^L \cos\theta(s)ds + \lambda_2 \int_0^L \sin\theta(s)ds + \lambda \int_0^L ds, \tag{4.108}$$

the Lagrangian of which, (taking into account that $\kappa(s) = \dfrac{d\theta(s)}{ds} = \dot\theta(s)$), is

$$F(\theta, \dot\theta, s) = \dot\theta^2(s) + \lambda_1 \cos\theta(s) + \lambda_2 \sin\theta(s) + \lambda. \tag{4.109}$$

The respective Euler–Lagrange equation is

$$\ddot\theta(s) + \frac{\lambda_1}{2}\sin\theta(s) - \frac{\lambda_2}{2}\cos\theta(s) = 0. \tag{4.110}$$

Differentiating both sides of the above equation with respect to s, we have

$$\dddot{\theta}(s) + \frac{\lambda_1}{2}\dot{\theta}(s)\cos\theta(s) + \frac{\lambda_2}{2}\dot{\theta}(s)\sin\theta(s) = 0. \tag{4.111}$$

On the other side, if we multiply both sides of (4.110) with $\dot{\theta}(s)$ and integrate, we have

$$\dot{\theta}^2(s) - \lambda_1\cos\theta(s) - \lambda_2\sin\theta(s) + 2\mu = 0, \tag{4.112}$$

where 2μ is the integration constant.

The proper combination of Eqs. (4.111) and (4.112) leads to the elimination of the Lagrange multipliers λ_1 and λ_2, and in this way, we get the equation

$$\ddot{\theta} + \frac{\dot{\theta}^3}{2} + \sigma\dot{\theta} = 0. \tag{4.113}$$

Since, by definition, $\dot{\theta} = \kappa$, we can aslo write

$$\ddot{\kappa} + \frac{\kappa^3}{2} + \mu\kappa = 0, \tag{4.114}$$

and see that this is nothing other than the intrinsic equation of the curves we are looking for. Further on, we will call it *Euler's elasticas equation*. The case in which $\mu \equiv 0$, i.e.,

$$\ddot{\kappa} + \frac{\kappa^3}{2} = 0 \tag{4.115}$$

is known in the literature as the *free elasticas equation*. One remarkable fact is that solutions to Eq. (4.114) can be derived in the same way as the deformation of the solutions to (4.115); in other words, the solutions to the free elasticas are sufficient to generate the Euler elasticas solutions. Bernoulli (1694) makes this very claim, but does not present any arguments. In a strictly analytical form, this is proved by Djondjorov et al. (2009).

It is also interesting to mention that under un appropriate choice of the coordinate system, the original equation of elasticas (4.110) can be written in the form

$$\ddot{\theta} + \lambda\sin\theta = 0, \tag{4.116}$$

which coincides with the equation of the mathematical pendulum (see Appendix A.1). The connection between the pendulunm and elasticas is shown in Fig. 4.13, and the explicit formulae (6.69), can be seen in Sect. 6.4.1.

Fig. 4.13 Examples of
conformity between Euler
elasticas and the motions of
the mathematical pendulum

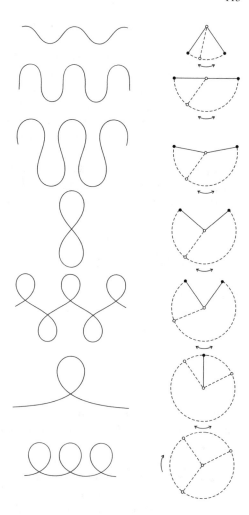

4.5   Whewell Parameterization

As has been pointed out, the problem of the curves describing the deflection of
elastic bars was investigated for the first time by Bernoulli (1694), who tried to apply
infinitesimal calculus, recently developed at the time, for solving the problem of
finding the respective shapes. The subsequent story can be traced back by looking at
recent review articles on the subject, cf. d'Antonio (2007), Goss (2009) and Matsutani
(2010).

Our interest in the subject is due to the amazing connections which appear to exist
between elasticas and mathematical models of axonemes (Ludu and Cibert 2009),
membranes (Vassilev et al. 2008) and nanotubes (Cox and Hill 2008). The guiding
idea of the latter paper is that the geometry of joints between a carbon nanotube and a

flat graphene sheet is a result of the minimization of the curvature of the profile curve of the axially symmetric neck which realizes the transition. In adjusting the solutions to the specific geometry of the nanotube and the graphene plane, however, a problem with boundary conditions appears, as the explicit elastica solutions (see Born 1906; Love 1944 or Djondjorov et al. 2008) are given in terms of the arclength parameter s, while the total length of the profile curve in most cases is unknown. Much more suitable in this situation from a geometrical point of view will be a parameterization written in terms of the slope of the tangent to the curve, the so-called (Whewell 1849, 1850) parameterization.

The main purpose of the section to follow is to provide the above-mentioned parameterization along with some clarifying remarks about the intrinsic equations of the free elastica and the elastica with tension and their solutions.

4.5.1 Equilibrium Equations

Following Euler (1744), one can derive the equilibrium equation of the elastica by considering the variational problem of minimizing the so-called bending energy $\mathcal{E}$ given by the integral of the squared curvature κ

$$\mathcal{E} = \int_{(\mathring{A},\mathring{\alpha})}^{(A,\alpha)} \kappa^2(s)\, ds \tag{4.117}$$

of the curve representing the thin beam passing through some fixed points $\mathring{A}$ and A on the plane and having there fixed slopes $\mathring{\alpha}$ and α to the OX axis. For the moment, we will confine our study to convex curves, i.e., to those curves the curvatures of which have fixed signs. Under these circumstances, the slope angles are monotonic functions of the respective arclength parameters.

We can also assume that the curvature radius of such a curve is positive and that its slope angle θ is an increasing function of s (if this is not fulfilled, it can be realized by a simple switch of the endpoints).

In these settings, we also have the fundamental geometrical relations

$$\frac{d\theta(s)}{ds} := \kappa(s), \qquad \kappa(s) > 0 \tag{4.118}$$

and

$$\frac{dx(s)}{ds} = \cos\theta(s), \qquad \frac{dz(s)}{ds} = \sin\theta(s), \tag{4.119}$$

from which one can easily obtain

$$\Delta x = \int_{\mathring{s}}^{s} \cos\theta(s)ds = \int_{\mathring{\alpha}}^{\alpha} \frac{\cos\theta}{\kappa(\theta)}d\theta \tag{4.120}$$

and

$$\Delta z = \int_{\dot{s}}^{s} \sin \theta(s) ds = \int_{\dot{\alpha}}^{\alpha} \frac{\sin \theta}{\kappa(\theta)} d\theta. \tag{4.121}$$

With some Lagrangian multipliers, say $\tilde{a}$ and $\tilde{c}$,

$$\mathcal{J} = \int \kappa(\theta) d\theta + \tilde{a} \int \frac{\cos \theta}{\kappa} d\theta + \tilde{c} \int \frac{\sin \theta}{\kappa} d\theta, \tag{4.122}$$

we have the Euler–Lagrange equation

$$\frac{\partial}{\partial \kappa} (\kappa + \tilde{a} \frac{\cos \theta}{\kappa} + \tilde{c} \frac{\sin \theta}{\kappa}) = 0,$$

which immediately produces

$$\kappa^2 = a \cos\theta + c \sin\theta, \qquad a = -\tilde{a}, \qquad c = -\tilde{c}, \tag{4.123}$$

and additionally, by a differentiation of both sides of (4.123) with respect to the natural parameter s, one also obtains

$$2\dot{\kappa} = c \cos\theta - a \sin\theta. \tag{4.124}$$

Squaring the left- and right-hand sides of (4.123) and (4.124) and adding them leads to the intrinsic equation of the so-called free elastica (see Birkhoff and de Boor 1965)

$$\dot{\kappa}^2(s) + \frac{\kappa^4(s)}{4} - \frac{m^2}{4} = 0, \tag{4.125}$$

in which $m^2 = a^2 + c^2$.

The fundamental theorem in classical differential geometry says that once the curvature of a plane curve is known, the curve can be reconstructed up to a position and orientation. Here, this can be seen by simply solving the system formed by (4.123) and (4.124) for $\cos\theta$ and $\sin\theta$, i.e.,

$$\cos\theta = \frac{1}{m^2}(a\kappa^2 + 2c\dot{\kappa}), \qquad \sin\theta = \frac{1}{m^2}(c\kappa^2 - 2a\dot{\kappa}), \tag{4.126}$$

and integrating Eqs. (4.119) to obtain the explicit expressions for the elastica line in the form

$$x = \frac{1}{m^2}(a\mathcal{E} + 2c\kappa), \qquad z = \frac{1}{m^2}(c\mathcal{E} - 2a\kappa), \tag{4.127}$$

in which the integration constants corresponding to the translations are omitted.

One way to carry this to the very end is to find $\kappa(s)$ by solving Eq. (4.125), and then evaluate the energy integral (4.117) as an explicit function of the arclength.

Another possibility is to make use of the Whewell parameterization, which is the route we are going to follow below. In this parameterization,

$$\kappa(\theta) = \sqrt{a\cos\theta + c\sin\theta} \quad \text{and therefore} \quad \mathcal{E} = \int \kappa(\theta)\, d\theta. \qquad (4.128)$$

Introducing

$$\psi(\theta) = \arcsin\sqrt{\frac{m - a\cos\theta - c\sin\theta}{m}}, \qquad \theta \in \left(\arcsin\frac{c}{m} - \frac{\pi}{2}, \arcsin\frac{c}{m}\right)$$

performing the integration in (4.128) and making use of (4.127), we end up with the following expressions:

$$x(\theta) = \frac{a\sqrt{2m}\left(2E(\psi(\theta), \frac{1}{\sqrt{2}}) - F(\psi(\theta), \frac{1}{\sqrt{2}})\right) + 2c\kappa(\theta)}{m^2}$$
$$\qquad\qquad\qquad\qquad\qquad\qquad\qquad\qquad\qquad\qquad (4.129)$$
$$z(\theta) = \frac{c\sqrt{2m}\left(2E(\psi(\theta), \frac{1}{\sqrt{2}}) - F(\psi(\theta), \frac{1}{\sqrt{2}})\right) - 2a\kappa(\theta)}{m^2}.$$

Going to another set of coordinates

$$\tilde{x} = ax + cz = \mathcal{E}, \qquad \tilde{z} = cx - az = 2\kappa, \qquad (4.130)$$

we actually get the parameterization of the rectangular elastica

$$\tilde{x}(\theta) = \sqrt{2m}\left(2E(\psi(\theta), \frac{1}{\sqrt{2}}) - F(\psi(\theta), \frac{1}{\sqrt{2}})\right), \quad \tilde{z}(\theta) = 2\sqrt{a\cos\theta + c\sin\theta}$$
$$\qquad\qquad\qquad\qquad\qquad\qquad\qquad\qquad\qquad\qquad (4.131)$$

that further on can be recognized as the profile curve of the so-called Mylar balloon of Mladenov and Oprea (2003a). The latter is depicted on the right-hand side of Fig. 4.14.

Fig. 4.14 Two elasticas shapes drawn, respectively, via formulas (4.129) (*left*) and (4.131) (*right*) for the same choice of the parameters $a = 1.2$ and $c = 0.2$

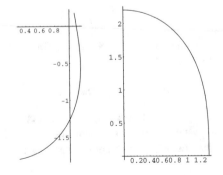

Remark 4.3 Inversion of (4.130), i.e., expressing x, z as (linear) functions of $\tilde{x}$ and $\tilde{z}$, leads to the conclusion that this transformation is actually a pseudo-rotation of the plane. This can be seen clearly in Fig. 4.14.

4.5.2 Elasticas With Tension

If we have to solve the energy minimization problem (4.117) under the constraint of the fixed length, i.e.,

$$\mathcal{L} = \int_{\tilde{\alpha}}^{\alpha} ds = \text{constant}, \tag{4.132}$$

the Lagrangian changes to

$$L := \mathcal{E} + \tilde{\sigma}\mathcal{L},$$

and the respective Euler–Lagrange equation turns out to be

$$\frac{\partial}{\partial \kappa}(\kappa + \tilde{a}\frac{\cos\theta}{\kappa} + \tilde{c}\frac{\sin\theta}{\kappa} + \frac{\tilde{\sigma}}{\kappa}) = 0,$$

where $\tilde{\sigma}$ is interpreted as the elastica's tension. In complete analogy with the free case, we have

$$\kappa^2 = a\cos\theta + c\sin\theta + \sigma \tag{4.133}$$

$$2\dot{\kappa} = c\cos\theta - a\sin\theta, \tag{4.134}$$

and the intrinsic equation this time becomes

$$\dot{\kappa}^2(s) + \frac{\kappa^4(s)}{4} - \frac{\sigma\kappa^2(s)}{2} + \frac{\sigma^2 - m^2}{4} = 0, \qquad m^2 = a^2 + c^2. \tag{4.135}$$

As before, the system formed by (4.133) and (4.134) can be solved for $\cos\theta$ and $\sin\theta$, i.e.,

$$\cos\theta = \frac{1}{m^2}(a(\kappa^2 - \sigma) + 2c\dot{\kappa}), \qquad \sin\theta = \frac{1}{m^2}(c(\kappa^2 - \sigma) - 2a\dot{\kappa}).$$

The explicit expressions for the elastica line are found to be

$$x_\sigma = \frac{1}{m^2}(a\mathcal{E} + 2c\kappa - a\sigma s), \qquad z_\sigma = \frac{1}{m^2}(c\mathcal{E} - 2a\kappa - c\sigma s). \tag{4.136}$$

Factorizing

$$-\kappa^4 + 2\sigma\kappa^2 + m^2 - \sigma^2,$$

which appears in (4.135), into the form

$$\left(\kappa^2 + m - \sigma\right)\left(m + \sigma - \kappa^2\right)$$

makes it obvious that the possible values of σ are either within the intervals $\ 0 <$ $|\sigma| < m, \quad 0 < m < \sigma$, or coincide with the point on the real line that separates them, i.e., $\sigma \equiv m$. In the first interval, the integral of the energy reads as

$$\mathcal{E}(\theta) = 2\sqrt{2m}\left(E(\phi(\theta), k) - \frac{m - \sigma}{2m}F(\phi(\theta), k)\right), \tag{4.137}$$

where

$$\phi(\theta) = \arcsin\sqrt{\frac{m - a\cos\theta - c\sin\theta}{m + \sigma}}, \qquad k = \sqrt{\frac{m + \sigma}{2m}}$$

$$\theta \in \left(\arccos\frac{a}{m} - \arccos(-\frac{\sigma}{m}), \ \arccos\frac{a}{m}\right), \tag{4.138}$$

and the arclength s is

$$s(\theta) = \int \frac{d\theta}{\kappa(\theta)} = \sqrt{\frac{2}{m}}F(\phi(\theta), k). \tag{4.139}$$

In the second interval, we have, respectively,

$$\mathcal{E}(\theta) = 2\sqrt{m + \sigma}E(\varphi(\theta), k), \qquad \varphi(\theta) = \arcsin\sqrt{\frac{m - a\cos\theta - c\sin\theta}{2m}}$$

$$k = \sqrt{\frac{2m}{m + \sigma}}, \qquad \theta \in \left(\arccos\frac{a}{m} - \pi, \ \arccos\frac{a}{m}\right) \tag{4.140}$$

$$s(\theta) = \int \frac{d\theta}{\kappa(\theta)} = \frac{2}{\sqrt{m + \sigma}}F(\varphi(\theta), k).$$

In the exceptional case $\sigma = m$, one can introduce the real parameter φ via $a = m\sin\varphi, c = m\cos\varphi$, so that $\kappa(\theta)$ becomes

$$\kappa(\theta) = \sqrt{2m}\cos\frac{\theta - \varphi}{2}, \tag{4.141}$$

which in turn produces

$$\mathcal{E}(\theta) = 2\sqrt{2m}\sin(\frac{\theta - \varphi}{2}), \qquad s(\theta) = 2\sqrt{\frac{2}{m}}\operatorname{arctanh}(\tan(\frac{\theta - \varphi}{2})). \tag{4.142}$$

Inserting the so-obtained expressions for $\mathcal{E}(\theta)$ and $s(\theta)$ into (4.136), one gets the sought-after elasticas arcs that can be concatenated appropriately if such need exists.

4.5.2.1 Remarks

The above considerations have some interesting consequences when applied to the study of elastic curves.

First of all, performing the same transformation as specified in (4.130), one gets the known formulas of the elastica

$$\tilde{x}_\sigma = a x_\sigma + c z_\sigma = \mathscr{E} - \sigma s$$
$$\tilde{z}_\sigma = c x_\sigma - a z_\sigma = 2\kappa.$$

One must notice, however, that this time, its parameterization is given in terms of the slope angle θ.

Next, in all considerations up to now, we have tacitly assumed that the curvature has a piecewise fixed sign. This is true for an arbitrary smooth curve as well, provided that it is presented as a union of convex arcs joining its inflection points. Within the i-th segment, the curvature satisfies Eqs. (4.133) and (4.134), i.e.,

$$\kappa_i^2 = a_i \cos\theta + c_i \sin\theta + \sigma_i \tag{4.143}$$
$$2\,\dot{\kappa}_i = c_i \cos\theta - a_i \sin\theta, \tag{4.144}$$

where a_i, c_i and σ_i are some constants. As the elastica is assumed to be homogeneous, it is clear that its tension is constant, and therefore $\sigma_i \equiv \sigma_{i+1} \equiv \cdots \equiv \sigma$. The continuity of the curvature and its derivatives at the inflections means that at each such point, one has the following system of equations:

$$\cos\theta_i(a_i - a_{i+1}) + \sin\theta_i(c_i - c_{i+1}) = 0$$
$$-\sin\theta_i(a_i - a_{i+1}) + \cos\theta_i(c_i - c_{i+1}) = 0, \tag{4.145}$$

in which θ_i is the slope angle of the tangent to the OX axis at the inflection point separating two consecutive segments, say, i and $i + 1$. As (4.145) is a linear system of non-singular homogeneous equations, it possesses a unique solution $a_i \equiv a_{i+1}$, $c_i \equiv c_{i+1}$, which is valid for all adjacent segments, and this means that (4.133) holds along the entire curve with the same a and c. It is also obvious that at each inflection point, the curvature changes its sign.

Furthermore, if the curve is specified and one has to know the constants a, c and σ, it is necessary to solve the nonlinear system formed by Eqs. (4.120), (4.121) and (4.132) (if the elastica length is fixed). For that purpose, use has to be made of the expression of the curvature either in the form (4.123) or (4.133), depending on the absence or presence of constraint.

Finally, using the Whewell parameterization for the elastica with tension derived in the previous section, it is easy to obtain all joints between a flat graphene sheet and carbon nanotubes described as Model I in Cox and Hill (2008), avoiding the problem with boundary conditions mentioned in the Introduction. Several plots of possible profile curves of such joints are depicted in Fig. 4.15.

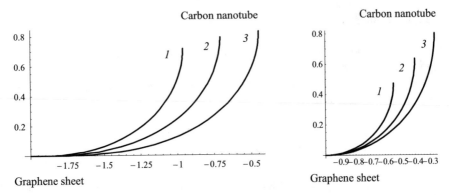

Fig. 4.15 Profile *curves* of possible joints between a flat graphene sheet and carbon nanotubes according to Cox's model drawn via (4.137)–(4.139) with parameters $a = c = \sqrt{2}$ and $\sigma = \sqrt{2}, 1.5, 1.782$

4.6 Elastic Sturmian Spirals

In this section, we study planar curves that simultaneously represent solutions to the Euler elastica problem and generalized Sturmian spirals. Let us start with a brief description of these physical concepts.

The elastic curve minimizes the integral of the curvature squared subject to fixed length and first order initial conditions. Let $\kappa(s)$ be the curvature of a curve parameterized by its arc length. We want to minimize the bending energy

$$\int_\gamma \kappa^2(s)\mathrm{d}s \tag{4.146}$$

among all curves γ which have a fixed length. As we know, after solving the variational problem, one gets that the curvature of such a planar curve satisfies the Euler–Lagrange equation

$$\ddot{\kappa} = -\frac{1}{2}\kappa^3 + \frac{\lambda}{2}\kappa, \tag{4.147}$$

where λ is the tension and the dots denote the derivatives with respect to the natural parameter s. An immediate integration of Eq. (4.147) yields

$$\dot{\kappa}^2 = -\frac{1}{4}\kappa^4 + \frac{\lambda}{2}\kappa^2 + 2E, \tag{4.148}$$

where E is the constant of integration (energy). This will be the equation of interest for us. For a more extensive description of elasticity phenomena, see Sect. 2.7 and (Singer 2008).

Next, we introduce the generalized Sturmian spirals (for more details, see Mladenov et al. 2011a). These are the planar curves with the property that at each point, their curvature radii are proportional to the distance from a fixed point in the plane. If we denote the positive constant multiplier by σ, all of the above means that

$$\kappa(r) = \frac{\sigma}{r} \tag{4.149}$$

is satisfied identically along the curve. If $\sigma = 1$, we get the classical Sturmian spiral. As mentioned, for more information on this topic, the reader is advised to have a look at Sect. 2.7 and (Mladenov et al., 2011a).

In order to determine these spirals, we will make use of the intrinsic equation of Euler's elasticas (4.148) and substitute their curvature (4.149) inside. In this way, we obtain the following equation:

$$\dot{r}^2 = \frac{2E}{\sigma^2}r^4 + \frac{\lambda}{2}r^2 - \frac{\sigma^2}{4} \equiv P(r). \tag{4.150}$$

Note that if we solve the above equation, we can easily reconstruct the curve. The reason is clear—if we know r, we also have κ as function of s, and then, by taking into account that

$$\frac{d\theta}{ds} = \kappa(s), \tag{4.151}$$

we can also find the inclination angle θ. Finally, relying on the Frenet frame (1.20), we can find the Cartesian coordinates by the formulas

$$x = \int \cos(\theta(s))ds, \quad z = \int \sin(\theta(s))ds. \tag{4.152}$$

Our goal will be to solve (4.150) for different $P(r)$, and therefore obtain expressions and pictures by (4.151) and (4.152).

4.6.1 Explicit Analytical Solutions

As discussed in the Introduction, we start with the intrinsic equation of the Eulerian elasticas, i.e., (4.148). Here, we assume that the tension λ, being a physical property, is a constant. Actually, there are two cases depending on the sign of E.
Case 1: We will first consider the case in which E and λ are both positive. We will write this fact as follows:

$$E = \frac{a^2 c^2}{8}, \quad \lambda = \frac{a^2 - c^2}{2}, \tag{4.153}$$

with $a^2 \geq c^2$. Let $\kappa = \sigma/r$. Thus Eq. (4.148) transforms into

$$\dot{r}^2 = \frac{2E}{\sigma^2}r^4 + \frac{\lambda}{2}r^2 - \frac{\sigma^2}{4}. \tag{4.154}$$

Using the formulae for E and λ, we get

$$\dot{r}^2 = \frac{2a^2c^2}{8\sigma^2}r^4 + \frac{(a^2 - c^2)}{4}r^2 - \frac{\sigma^2}{4}$$
$$\dot{r}^2 = \frac{a^2c^2}{4\sigma^2}\left(r^4 + \frac{(a^2 - c^2)\sigma^2}{a^2c^2}r^2 - \frac{\sigma^4}{a^2c^2}\right) \equiv \frac{a^2c^2}{4\sigma^2}A(r), \tag{4.155}$$

where the expression $A(r)$ in the large brackets is a quartic polynomial. This polynomial splits nicely, and we get

$$A(r) = \left(r^2 - \frac{\sigma^2}{a^2}\right)\left(r^2 + \frac{\sigma^2}{c^2}\right), \tag{4.156}$$

and therefore

$$\dot{r} = \frac{ac}{2\sigma}\sqrt{\left(r^2 - \frac{\sigma^2}{a^2}\right)\left(r^2 + \frac{\sigma^2}{c^2}\right)} \tag{4.157}$$

$$r = \frac{\sigma}{a}\frac{1}{\mathrm{cn}(\frac{\sqrt{a^2+c^2}}{2}s, k)}, \qquad k = \frac{a}{\sqrt{a^2 + c^2}}, \qquad r > \frac{\sigma}{a}. \tag{4.158}$$

According to Vassilev et al. (2008), the Cartesian coordinates x, z corresponding to the solution of Eq. (4.158), are

$$x(s) = C \int \frac{\sigma^2 ds}{r^2(s)} - \frac{a^2 - c^2}{a^2 + c^2}s, \qquad z(s) = -\frac{2\sigma C}{r}, \tag{4.159}$$

with C a constant $C = \frac{1}{\kappa^2(0) - \lambda} = \frac{2}{a^2 + c^2}$. It is interesting that one of the components, z reflects the spiral property, while the other, x, does the same for the elasticity of the curve. Continuing with the solution of the first case presented in (4.158) and (4.159), we get

$$x(u) = \frac{4a^2}{(a^2 + c^2)^{3/2}}\int \mathrm{cn}^2(u)du - \frac{a^2 - c^2}{a^2 + c^2}s \tag{4.160}$$
$$z(u) = -\frac{4a}{a^2 + c^2}\mathrm{cn}(\frac{\sqrt{a^2 + c^2}}{2}s, k).$$

By taking the integral for $x(u)$ (see Byrd and Friedman 1954), we produce the parameterization

$$x(u) = \frac{4a^2}{k^2(a^2+c^2)^{3/2}}(E(\text{am}(\frac{\sqrt{a^2+c^2}}{2}s,k),k) - \tilde{k}^2\frac{\sqrt{a^2+c^2}}{2}s) - \frac{a^2-c^2}{a^2+c^2}s$$

$$z(u) = -\frac{4a}{a^2+c^2}\text{cn}(\frac{\sqrt{a^2+c^2}}{2}s,k). \tag{4.161}$$

Here, $E(\text{am}(v,k),k)$ is the elliptic integral of the second kind and, as usual k denotes the elliptic modulus, while $\tilde{k}$ is the so-called complementary elliptic modulus $(k^2+\tilde{k}^2=1)$. Taking into account that

$$k = \frac{a}{\sqrt{a^2+c^2}}, \qquad \tilde{k} = \frac{c}{\sqrt{a^2+c^2}},$$

the expressions in (4.161) turn into

$$x(u) = \frac{4}{\sqrt{a^2+c^2}}E(\text{am}(\frac{\sqrt{a^2+c^2}}{2}s,k),k) - s \tag{4.162}$$

$$z(u) = -\frac{4a}{a^2+c^2}\text{cn}(\frac{\sqrt{a^2+c^2}}{2}s,k).$$

The plots in Fig. 4.16 present two examples of use of the formulas in (4.162). We get similar graphs for other values of the parameters. Of course, one needs to shift these graphs so that the Sturmian spiral property can be seen immediately. Also, one needs to use appropriate bounds for the arclengths. Nevertheless, these are pieces of Sturmian spirals that are also elastic curves (Figs. 4.16 and 4.17).

Case 2: This is the case in which the energy E is negative, and according to (Vassilev et al., 2008), this means that one can write

$$E = -\frac{a^2c^2}{8}, \qquad \lambda = \frac{a^2+c^2}{2}.$$

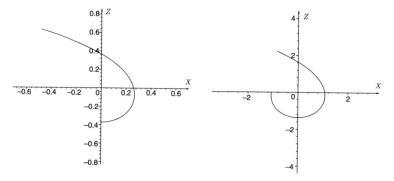

Fig. 4.16 Plot of (4.162) with $a=4, c=1$ (*left*) and $a=1, c=0.3$ (*right*)

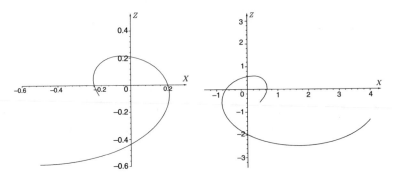

Fig. 4.17 Plot of (4.166) with $a = 1, c = 4$ (*left*) and $a = 0.3, c = 1$ (*right*)

Respectively, we have as equation and solution

$$\dot{r} = \frac{ac}{2\sigma}\sqrt{\left(\frac{\sigma^2}{a^2} - r^2\right)\left(r^2 - \frac{\sigma^2}{c^2}\right)}$$

$$r = \frac{\sigma}{a}\mathrm{dn}(\frac{c}{2}s, k), \qquad k = \frac{\sqrt{c^2 - a^2}}{c}, \qquad \frac{\sigma}{a} > r > \frac{\sigma}{c}. \qquad (4.163)$$

This case can be treated in the same way as the previous one, and in the end, one obtains

$$x(u) = \frac{4a^2}{c(a^2 - c^2)}\int \mathrm{nd}^2(u)du - \frac{a^2 + c^2}{a^2 - c^2}s \qquad (4.164)$$

$$z(u) = \frac{4a}{c^2 - a^2}\mathrm{nd}(\frac{c}{2}s, k),$$

which, after integration, becomes

$$x(u) = \frac{4a^2}{ck^2(a^2 - c^2)}(E(\mathrm{am}(\frac{c}{2}s, k), k) - k^2\mathrm{sn}(\frac{c}{2}s, k)\mathrm{cd}(\frac{c}{2}s, k)) - \frac{a^2 + c^2}{a^2 - c^2}s \qquad (4.165)$$

$$z(u) = \frac{4a}{c^2 - a^2}\mathrm{nd}(\frac{c}{2}s, k).$$

Some further simplifications lead to

$$x(u) = -\frac{4c}{c^2 - a^2}E(\mathrm{am}(\frac{c}{2}s, k), k) + \frac{4}{c}\mathrm{sn}(\frac{c}{2}s, k)\mathrm{cd}(\frac{c}{2}s, k) + \frac{a^2 + c^2}{c^2 - a^2}s$$

$$z(u) = \frac{4a}{c^2 - a^2}\mathrm{nd}(\frac{c}{2}s, k). \qquad (4.166)$$

Fig. 4.18 Plot of an elastic spiral with $\mathcal{E} > 0$

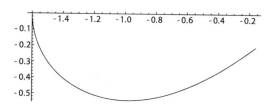

The figures above depict particular solutions for **Case** 2, similarly shifted as those in Fig. 4.18. Indeed, they look like pieces of spirals. It appears that the parameter σ of the spiral does not contribute much to the pictures, as neither the coordinates nor the curvatures depend on it explicitly.

4.7 Alternative Parameterization of Elastic Spirals

Instead of having explicit solutions to Eq. (4.150), we can try another approach. Let us start by noticing that Eq. (4.150) gives us the relation $ds = dr/\sqrt{P(r)}$. Using (4.149) and replacing in (4.151), we have $d\theta = \sigma ds/r$. Combining both relations, we end up with the equation

$$d\theta = \frac{\sigma dr}{r\sqrt{P(r)}}, \tag{4.167}$$

which gives us the connection between the polar radius and the angle between the tangent vector and the horizontal axis. As $P(r)$ is a fourth order polynomial without odd degree terms, we are able to solve (4.167). Since we will look for spiral curves, we can do the integration in (4.152) with respect to r. Typically for a spiral, r will increase with s, and thus we can write

$$x = \int \cos(\theta(s(r))) \frac{ds}{dr} dr, \quad z = \int \sin(\theta(s(r))) \frac{ds}{dr} dr. \tag{4.168}$$

Integrating with respect to r seems more natural for curves of this type and appears to be computationally less expensive.

4.7.1 Main Calculations

In this section, we will assume throughout, that the tension λ is positive. We will use relation (4.167) to directly attack the problem. To do this, one can look at the formula in the first case presented in Marinov et al. (2014).

4.7.1.1 Positive Energy Case

First, we work out the positive energy case. Here, we assume

$$\mathcal{E} = \frac{a^2 c^2}{8}, \qquad \lambda = \frac{a^2 - c^2}{2} > 0. \tag{4.169}$$

We use the afore-mentioned formula from Marinov et al. (2014) to transforms (4.150) by substituting the energy $\mathcal{E}$ and the tension λ with the new parameters a and c. We thus have

$$\dot{r} = \frac{ac}{2\sigma} \sqrt{\left(r^2 - \frac{\sigma^2}{a^2}\right)\left(r^2 + \frac{\sigma^2}{c^2}\right)}, \tag{4.170}$$

and by changing the variables with θ, we get

$$\frac{d\theta}{ds} = \kappa = \frac{\sigma}{r}, \qquad \frac{dr}{ds} = \frac{dr}{d\theta}\frac{d\theta}{ds},$$

so therefore

$$d\theta = \frac{2\sigma^2 dr}{acr\sqrt{\left(r^2 - \frac{\sigma^2}{a^2}\right)\left(r^2 + \frac{\sigma^2}{c^2}\right)}}. \tag{4.171}$$

Making the appropriate replacements $\tilde{a} = \sigma/a$ and $\tilde{c} = \sigma/c$, we have

$$d\theta = \frac{2\tilde{a}\tilde{c}dr}{r\sqrt{\left(r^2 - \tilde{a}^2\right)\left(r^2 + \tilde{c}^2\right)}}, \tag{4.172}$$

which integrates to

$$\theta + C = \tan^{-1}\left(\frac{(\tilde{c}^2 - \tilde{a}^2)r^2 - 2\tilde{a}^2\tilde{c}^2}{2\tilde{a}\tilde{c}\sqrt{\left(r^2 - \tilde{a}^2\right)\left(r^2 + \tilde{c}^2\right)}}\right) \equiv \tan^{-1}(W). \tag{4.173}$$

Setting $C = 0$ in the above formula gives us θ as a function of $(r, \tilde{a}, \tilde{c})$, or alternatively, using the original parameters, the expressions $\theta = \theta(r, a, c, \sigma)$

$$\cos(\theta) = \cos(\tan^{-1}(W)) = \frac{1}{\sqrt{1 + W^2}} = \frac{2\tilde{a}\tilde{c}\sqrt{\left(r^2 - \tilde{a}^2\right)\left(r^2 + \tilde{c}^2\right)}}{(\tilde{a}^2 + \tilde{c}^2)r^2} \tag{4.174}$$

$$\sin(\theta) = \sin(\tan^{-1}(W)) = \frac{W}{\sqrt{1 + W^2}} = \frac{(\tilde{c}^2 - \tilde{a}^2)r^2 - 2\tilde{a}^2\tilde{c}^2}{(\tilde{a}^2 + \tilde{c}^2)r^2}.$$

Finally, since

$$\frac{ds}{dr} = \frac{2\tilde{a}\tilde{c}}{\sigma\sqrt{\left(r^2 - \tilde{a}^2\right)\left(r^2 + \tilde{c}^2\right)}},$$

we get formulae for $\cos(\theta(r))ds/dr$ and $\sin(\theta(r))ds/dr$ as in (4.168), to proceed with integration. The one with the cosine will depend only on r^2 and will integrate to a scalar multiple of the curvature, which reflects the spiral property of the solution curve. The other coordinate is more related to the elastic nature, since it has a term which is a scalar multiple of the bending energy. Next, we get

$$x = -\frac{4\tilde{a}^2\tilde{c}^2}{\sigma(\tilde{a}^2 + \tilde{c}^2)r} = -\frac{4\sigma}{(a^2 + c^2)r} = -\frac{4\kappa}{(a^2 + c^2)}$$

$$z = \int \frac{2\tilde{a}\tilde{c}((\tilde{c}^2 - \tilde{a}^2)r^2 - 2\tilde{a}^2\tilde{c}^2)}{\sigma(\tilde{a}^2 + \tilde{c}^2)r^2\sqrt{\left(r^2 - \tilde{a}^2\right)\left(r^2 + \tilde{c}^2\right)}} dr. \tag{4.175}$$

We can compute the integral for z. Using a formula in Gradshteyn and Ryzhik (2007), we get

$$z(r) = \frac{2\tilde{a}\tilde{c}(\tilde{c}^2 - \tilde{a}^2)}{\sigma(\tilde{a}^2 + \tilde{c}^2)^{\frac{3}{2}}} F\left(\cos^{-1}\left(\frac{\tilde{a}}{r}\right), \frac{\tilde{a}}{\sqrt{\tilde{a}^2 + \tilde{c}^2}}\right)$$

$$- \frac{4\tilde{a}^3\tilde{c}^3}{\sigma(\tilde{a}^2 + \tilde{c}^2)} \int \frac{dr}{r^2\sqrt{\left(r^2 - \tilde{a}^2\right)\left(r^2 + \tilde{c}^2\right)}}.$$

Going back to the original parameters, we have

$$z(r) = \frac{2(a^2 - c^2)}{(a^2 + c^2)^{\frac{3}{2}}} F\left(\cos^{-1}\left(\frac{\sigma}{ar}\right), \frac{a}{\sqrt{a^2 + c^2}}\right)$$

$$- \frac{4\sigma^3}{ac(a^2 + c^2)} \int \frac{dr}{r^2\sqrt{\left(r^2 - a^2\right)\left(r^2 + c^2\right)}} \tag{4.176}$$

$$= \left(\frac{2(a^2 - c^2)}{(a^2 + c^2)^{\frac{3}{2}}} + \frac{4c^2}{(a^2 + c^2)^{\frac{3}{2}}}\right) F\left(\cos^{-1}\left(\frac{\sigma}{ar}\right), \frac{a}{\sqrt{a^2 + c^2}}\right)$$

$$- \frac{4}{\sqrt{a^2 + c^2}} E\left(\cos^{-1}\left(\frac{\sigma}{ar}\right), \frac{a}{\sqrt{a^2 + c^2}}\right),$$

and finally,

$$x(r) = -\frac{4\tilde{a}^2\tilde{c}^2}{\sigma(\tilde{a}^2+\tilde{c}^2)r} = -\frac{4\sigma}{(a^2+c^2)r} = -\frac{4\kappa}{(a^2+c^2)} = -\frac{2\kappa}{\sqrt{\lambda^2+8\mathcal{E}}}$$

$$z(r) = \frac{2}{\sqrt{a^2+c^2}}F\left(\cos^{-1}\left(\frac{\sigma}{ar}\right), \frac{a}{\sqrt{a^2+c^2}}\right) \qquad (4.177)$$

$$-\frac{4}{\sqrt{a^2+c^2}}E\left(\cos^{-1}\left(\frac{\sigma}{ar}\right), \frac{a}{\sqrt{a^2+c^2}}\right).$$

These expressions are in terms of the incomplete elliptic integrals of the first and second kinds, F and E, respectively. The second argument of these functions is the elliptic modulus. We have now obtained a formula for the coordinates of the elastic spiral in terms of r, when $\mathcal{E}$, λ are both positive.

Figure 4.18 depicts the case in which $a = 2$ and $c = 1$. It does indeed resemble a piece of a spiral.

Next is the case in which the energy $\mathcal{E}$ is negative (also explored in (Marinov et al., 2014)).

4.7.1.2 Negative Energy Case

In the case in which the energy is negative, one can choose new parameters so that

$$\mathcal{E} = -\frac{a^2c^2}{8}, \qquad \lambda = \frac{a^2+c^2}{2},$$

with $c > a$. According to Marinov et al. (2014), we have

$$\dot{r} = \frac{ac}{2\sigma}\sqrt{\left(\frac{\sigma^2}{a^2}-r^2\right)\left(r^2-\frac{\sigma^2}{c^2}\right)}. \qquad (4.178)$$

Similarly to (4.171), we also have

$$d\theta = \frac{2\sigma^2 dr}{acr\sqrt{\left(\frac{\sigma^2}{a^2}-r^2\right)\left(r^2-\frac{\sigma^2}{c^2}\right)}}, \qquad (4.179)$$

or equivalently, again using $\tilde{a} \equiv \frac{\sigma}{a} > r > \frac{\sigma}{c} \equiv \tilde{c}$

$$d\theta = \frac{2\tilde{a}\tilde{c}dr}{r\sqrt{\left(\tilde{a}^2-r^2\right)\left(r^2-\tilde{c}^2\right)}}, \qquad (4.180)$$

we end up with

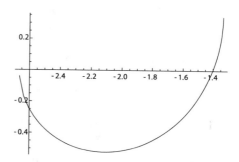

Fig. 4.19 Plot of an elastic spiral with $\mathcal{E} < 0$

$$\theta + C = \tan^{-1}\left(\frac{(\tilde{a}^2 + \tilde{c}^2)r^2 - 2\tilde{a}^2\tilde{c}^2}{2\tilde{a}\tilde{c}\sqrt{(\tilde{a}^2 - r^2)(r^2 - \tilde{c}^2)}} \right) \equiv \tan^{-1}(W). \tag{4.181}$$

Ultimately, we will use (4.168) as before. Calculations similar to those in the first case of this section lead to similar results, namely,

$$x(r) = -\frac{4\tilde{a}^2\tilde{c}^2}{\sigma(\tilde{a}^2 - \tilde{c}^2)r} = \frac{4\sigma}{(a^2 - c^2)r} = \frac{4\kappa}{(a^2 - c^2)} = -\frac{2\kappa}{\sqrt{\lambda^2 + 8E}}$$

$$z(r) = \frac{2(c^2 + a^2)}{c(c^2 - a^2)} F\left(\sin^{-1}\left(\frac{1}{r}\sqrt{\frac{c^2r^2 - \sigma^2}{c^2 - a^2}} \right), \frac{\sqrt{c^2 - a^2}}{c} \right) \tag{4.182}$$

$$-\frac{4c}{c^2 - a^2} E\left(\sin^{-1}\left(\frac{1}{r}\sqrt{\frac{c^2r^2 - \sigma^2}{c^2 - a^2}} \right), \frac{\sqrt{c^2 - a^2}}{c} \right).$$

Notice that for both cases, z can be easily given as a function of x, i.e., we have a local formula for a graph. Naturally, the expression for x in terms of λ and $\mathcal{E}$ agrees with the first case, and the next section as well.

Figure 4.19 represents the case with negative total energy with $c = 2$ and $a = 1$. Again, it looks like a piece of a spiral. We continue with the last case to which we dedicate a separate section.

4.7.1.3 Zero Tension

For completeness, we will provide another example that is relatively easier than the general case. Let $\mathcal{E} > 0$, $\sigma = 1$ and $\lambda = 0$, i.e., $P(r) = 2\mathcal{E}r^4 - 1/4$. In other words, there is no tension, and we will be looking for a classical Sturmian spiral. The energy $\mathcal{E}$ must be positive, since otherwise the expression under the radical in (4.183) will be negative as $\lambda = 0$. If we choose the constant of integration to be zero, the solution to (4.167) becomes

$$\theta = \int \frac{2dr}{r\sqrt{8\mathcal{E}r^4 - 1}} = -\cot^{-1}(\sqrt{8\mathcal{E}r^4 - 1}), \tag{4.183}$$

and this relation can be easily reduced to

$$\sqrt{8\mathcal{E}}r^2 \sin(\theta) + 1 = 0. \tag{4.184}$$

Let us check that (4.184) indeed represents a Sturmian spiral. By differentiating it with respect to s, we get

$$\cos(\theta)\kappa = \frac{1}{\sqrt{2\mathcal{E}}r^3}\frac{dr}{ds}$$

or

$$\frac{\sqrt{8\mathcal{E}r^4 - 1}}{\sqrt{8\mathcal{E}}r^2}\kappa = \frac{1}{\sqrt{2\mathcal{E}}r^3}\frac{\sqrt{8\mathcal{E}r^4 - 1}}{2},$$

which proves that $\kappa = 1/r$. Remember that we have a spiral, so r increases with s, and therefore the implicit function theorem applies. Now, if we invert the arc length and the curvature as $s = s(r)$ (which is precisely formula (4.168)), formula (4.152) becomes

$$x = \int \cos(\theta(r))\frac{ds}{dr}dr, \qquad y = \int \sin(\theta(r))\frac{ds}{dr}dr, \tag{4.185}$$

and furthermore, $ds/dr = 1/\sqrt{P(r)} = 2/\sqrt{8\mathcal{E}r^4 - 1}$. Finally, formula (4.185) becomes

$$x = \int \cos(-\cot^{-1}(\sqrt{8\mathcal{E}r^4 - 1}))\frac{2dr}{\sqrt{8\mathcal{E}r^4 - 1}} = \int \frac{\sqrt{8\mathcal{E}r^4 - 1}}{\sqrt{8\mathcal{E}}r^2}\frac{2dr}{\sqrt{8\mathcal{E}r^4 - 1}}$$

$$= -\frac{1}{\sqrt{2\mathcal{E}}r} + c_1 = -\frac{\kappa}{\sqrt{2\mathcal{E}}} + c_1 \tag{4.186}$$

$$z = \int \sin(-\cot^{-1}(\sqrt{8\mathcal{E}r^4 - 1}))\frac{2dr}{\sqrt{8\mathcal{E}r^4 - 1}} = -\frac{1}{\sqrt{2\mathcal{E}}}\int \frac{dr}{r^2\sqrt{8\mathcal{E}r^4 - 1}}$$

$$= -\frac{1}{\sqrt{8\mathcal{E}}}\int \frac{d\theta}{r} = -\frac{1}{\sqrt{8\mathcal{E}}}\int \kappa^2(s)ds.$$

Formula (4.186) reads such that we can parametrize the curve with negative multiples of its curvature and bending energy (4.146). As in the previous section, one of the coordinates resembles the elastic property, and the other its spiral nature. We can integrate the second part of (4.186) using elliptic functions to get an explicit solution. For that purpose, we substitute $(8\mathcal{E})^{1/4}r$ with u and perform a direct integration (cf Gradshteyn and Ryzhik 2007)

$$z = -\frac{1}{\sqrt{2\varepsilon}}\int\frac{dr}{r^2\sqrt{8\varepsilon r^4-1}} = -\frac{\sqrt{8\varepsilon r^4-1}}{\sqrt{2\varepsilon}r} + 2\sqrt{8\varepsilon}\int\frac{r^2dr}{\sqrt{8\varepsilon r^4-1}}$$

$$= -\frac{\sqrt{8\varepsilon r^4-1}}{\sqrt{2\varepsilon}r} + \frac{2}{(8\varepsilon)^{1/4}}\int\frac{u^2du}{\sqrt{u^4-1}}$$

$$= -\frac{\sqrt{8\varepsilon r^4-1}}{\sqrt{2\varepsilon}r} + \frac{\sqrt{2}}{(8\varepsilon)^{1/4}}F\left(\cos^{-1}\left(\frac{1}{u}\right),\frac{1}{\sqrt{2}}\right)$$

$$-\frac{2\sqrt{2}}{(8\varepsilon)^{1/4}}E\left(\cos^{-1}\left(\frac{1}{u}\right),\frac{1}{\sqrt{2}}\right) + \frac{2}{(8\varepsilon)^{1/4}}\frac{\sqrt{u^4-1}}{u}$$

$$= \frac{1}{(2\varepsilon)^{1/4}}F\left(\cos^{-1}\left(\frac{1}{u}\right),\frac{1}{\sqrt{2}}\right) - \frac{2}{(2\varepsilon)^{1/4}}E\left(\cos^{-1}\left(\frac{1}{u}\right),\frac{1}{\sqrt{2}}\right)$$

$$= \frac{1}{(2\varepsilon)^{1/4}}F\left(\cos^{-1}(\frac{1}{(8\varepsilon)^{1/4}r}),\frac{1}{\sqrt{2}}\right) - \frac{2}{(2\varepsilon)^{1/4}}E\left(\cos^{-1}(\frac{1}{(8\varepsilon)^{1/4}r}),\frac{1}{\sqrt{2}}\right). \quad (4.187)$$

The formula above agrees completely with the corresponding one in the previous section.

As an illustration, let us give an example in which $\mathcal{E} = 1/2$ with zero λ and $\sigma = 1$. In this case, r should be bounded below from $1/\sqrt{2}$, so there is no problem that it is reciprocal to x. The parameterization in this case looks quite simple and elegant. According to (4.187), we will have

$$x = -\frac{1}{r}$$
$$z = F\left(\cos^{-1}\left(\frac{1}{\sqrt{2}r}\right),\frac{1}{\sqrt{2}}\right) - 2E\left(\cos^{-1}\left(\frac{1}{\sqrt{2}r}\right),\frac{1}{\sqrt{2}}\right), \quad (4.188)$$

or locally, as a function $z = z(x)$

$$z(x) = F\left(\cos^{-1}\left(-\frac{x}{\sqrt{2}}\right),\frac{1}{\sqrt{2}}\right) - 2E\left(\cos^{-1}\left(-\frac{x}{\sqrt{2}}\right),\frac{1}{\sqrt{2}}\right). \quad (4.189)$$

Figure 4.20 is the graph of formula (4.189) and hints once again a both of the desired properties of the solution—the elastic and the spiral. Actually, this is another parameterization of the rectangular elastica which appears as the profile curve of the mylar balloon in Hadzhilazova and Mladenov (2008).

Fig. 4.20 Plot of an elastic spiral with a zero tension ($\lambda = 0$), and $\mathcal{E} = 1/2$, $\sigma = 1$

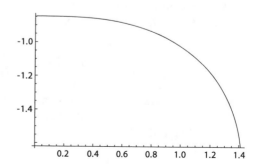

4.7.1.4   Remarks

To summarize the results, we give a new natural parameterizations of the generalized elastic Sturmian spirals. Closed formulae are derived. The solutions are given in terms of elliptic integrals depending on the polar radius and the physical parameters of the system.

We would like to mention that for the problem of finding elastic spirals, the Sturmian were not selected by accident. This is because when they get substituted into the Euler–Lagrange equation (4.148), one gets almost the same equation just in terms of the radius. This is guaranteed by the straightforward relation between radius and curvature and by the bi-quadratic nature of (4.148). Of course, if another class of spirals is used, the transition from (4.148) to (4.150) can add an odd degree or even more complicated terms in $P(r)$. Nevertheless, the results above are a solid base for further investigations into curves with such a dual physical nature.

4.8   Geometry of the Rotating Liquid Drop

Here, we consider the problem of a fluid body rotating with a constant angular velocity and subjected to surface tension. Determining the equilibrium configuration of this system turns out to be equivalent to the geometrical problem of determining the surface of revolution with a prescribed mean curvature.

In the simply connected case, the equilibrium surface can be parameterized explicitly via elliptic integrals (cf Jahnke et al. 1960) of the first and second kinds.

Here, we present two such parameterizations of the drops, and we use the second of them to study the finer details of the drop surfaces, such as the existence of closed geodesics (Oprea 2007).

4.8.1   Introduction

Let us consider a fluid body with an incompressible mass rotating with a constant angular velocity and subjected to surface tension. The problem then is to find the shape which this mass will have under the prescribed angular velocity.

The energy of the drop enclosing a volume V is given by

$$\mathcal{E} = \sigma \mathcal{S} - \frac{\rho \omega^2}{2} \int_V r^2 \mathrm{d}r, \tag{4.190}$$

where σ is the surface tension, $\mathcal{S}$ is the area of the free surface, ρ is the fluid density, ω is the fixed angular velocity and r is the radial distance from the axis of rotation. The corresponding Euler–Lagrange equation for the equilibrium configuration reduces to the equation

$$H = 2\tilde{a}r^2 + \tilde{c}, \tag{4.191}$$

in which H is the mean curvature of $\mathcal{S}$, $\tilde{a}$ is a positive constant and the constant $\tilde{c}$ arises from the volume constraint.

4.8.2 Geometry and Surface Invariants

Let us now recall the fundamental relations among meridional curvature κ_μ, circumferential curvature κ_π and the mean H curvature for surfaces of revolution

$$\kappa_\mu = \frac{d(r\kappa_\pi)}{dr}, \qquad H = \frac{\kappa_\mu + \kappa_\pi}{2}. \tag{4.192}$$

The solution to this system of equations is

$$\kappa_\pi = \tilde{a}r^2 + \tilde{c} + \frac{C}{r^2}, \qquad \kappa_\mu = 3\tilde{a}r^2 + \tilde{c} - \frac{C}{r^2}, \tag{4.193}$$

where C is an integration constant which will be assumed to be zero from now on.

Moreover, if $\mathcal{R}_\pi$ is the distance from the point on the surface to the intersection of the normal at that point with the symmetry axes (see Fig. 4.21), then we additionally have

$$\kappa_\pi = \frac{1}{\mathcal{R}_\pi} = \frac{\sin\theta}{r}, \tag{4.194}$$

and therefore

$$\frac{dz}{dr} = -\tan\theta = -\frac{r\kappa_\pi}{\sqrt{1 - (r\kappa_\pi)^2}}. \tag{4.195}$$

Finally, the profile curve is given by the integral

Fig. 4.21 Geometry of the
profile *curve*

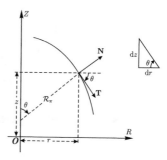

$$z(r) = -\int \frac{(\tilde{a}r^3 + \tilde{c}r)dr}{\sqrt{1 - (\tilde{a}r^3 + \tilde{c}r)^2}}. \qquad (4.196)$$

Having this result, we can compute, respectively, the coefficients E, F, G of the first and L, M, N of the second fundamental form of $\mathcal{S}$ (see the Maple work sheet below)

$$E = \frac{1}{1 - (\tilde{a}r^3 + \tilde{c}r)^2}, \qquad F = 0, \qquad G = r^2 \qquad (4.197)$$

$$L = \frac{3\tilde{a}r^2 + \tilde{c}}{1 - (\tilde{a}r^3 + \tilde{c}r)^2}, \qquad M = 0, \qquad N = r^2(\tilde{a}r^2 + \tilde{c}).$$

The following procedures help to calculate the surface invariants of the drop, in particular, its Gaussian and mean curvatures. (Warning: in the Maple code which follows, u plays the role of r and the tildas over a and c were omitted for easy encoding, but should be understood by the reader.)

```
>   EFG := proc(X)
>   local Xu,Xv,E,F,G;
>   Xu := <diff(X[1],u),diff(X[2],u),diff(X[3],u)>;
>   Xv := <diff(X[1],v),diff(X[2],v),diff(X[3],v)>;
>   E := DotProduct(Xu,Xu,conjugate=false);
>   F := DotProduct(Xu,Xv,conjugate=false);
>   G := DotProduct(Xv,Xv,conjugate=false);
>   simplify([E,F,G]);
>   end:
>   UN := proc(X)
>   local Xu,Xv,Z,s;
>   Xu := <diff(X[1],u),diff(X[2],u),diff(X[3],u)>;
>   Xv := <diff(X[1],v),diff(X[2],v),diff(X[3],v)>;
>   Z := -CrossProduct(Xu,Xv);
>   s:=VectorNorm(Z,Euclidean,conjugate=false);
>   simplify(<Z[1]/s|Z[2]/s|Z[3]/s>,sqrt,trig,symbolic);
>   end:
>   lmn := proc(X)
>   local Xu,Xv,Xuu,Xuv,Xvv,U,l,m,n;
>   Xu := <diff(X[1],u),diff(X[2],u),diff(X[3],u)>;
>   Xv := <diff(X[1],v),diff(X[2],v),diff(X[3],v)>;
>   Xuu := <diff(Xu[1],u),diff(Xu[2],u),diff(Xu[3],u)>;
>   Xuv := <diff(Xu[1],v),diff(Xu[2],v),diff(Xu[3],v)>;
>   Xvv := <diff(Xv[1],v),diff(Xv[2],v),diff(Xv[3],v)>;
>   U := UN(X);
>   l := DotProduct(U,Xuu,conjugate=false);
>   m := DotProduct(U,Xuv,conjugate=false);
>   n := DotProduct(U,Xvv,conjugate=false);
>   simplify([l,m,n],sqrt,trig,symbolic);
>   end:
```

Finally, we can calculate the Gauss curvature K as follows:

```
>   GK := proc(X)
>   local E,F,G,l,m,n,S,T;
>   S := EFG(X);
>   T := lmn(X);
>   E := S[1];
>   F := S[2];
>   G := S[3];
>   l := T[1];
>   m := T[2];
>   n := T[3];
>   simplify((l*n-m^2)/(E*G-F^2),sqrt,trig,symbolic);
>   end:
```

Mean curvature is given by

```
>   MK := proc(X)
>   local E,F,G,l,m,n,S,T;
>   S := EFG(X);
>   T := lmn(X);
>   E := S[1];
>   F := S[2];
>   G := S[3];
>   l := T[1];
>   m := T[2];
>   n := T[3];
>   simplify((G*l+E*n-2*F*m)/(2*E*G-2*F^2),sqrt,trig,symbolic);
>   end:
```

Now let us use these procedures to calculate curvatures for the drop using the first parameterization we gave.

```
>   XX:=<u*cos(v),u*sin(v),Int((a*u^3+c*u)/sqrt(1-(a*u^3+c*u)^2),
>   u)>;
```

$$XX := \left[\begin{array}{c} u\cos(v) \\ u\sin(v) \\ \displaystyle\int \frac{a\,u^3 + c\,u}{\sqrt{1 - a^2\,u^6 - 2\,a\,u^4\,c - c^2\,u^2}}\,du \end{array} \right]$$

```
>   EFG(XX);
```

$$[-\frac{1}{-1 + a^2\,u^6 + 2\,a\,u^4\,c + c^2\,u^2},\ 0,\ u^2]$$

```
>   simplify(lmn(XX),symbolic);
```

$$[-\frac{3\,a\,u^2 + c}{-1 + a^2\,u^6 + 2\,a\,u^4\,c + c^2\,u^2},\ 0,\ u^2\,(a\,u^2 + c)]$$

```
>   simplify(MK(XX),symbolic);
```

$$2\,a\,u^2 + c$$

```
>   GK(XX);
```

$$(3\,a\,u^2 + c)\,(a\,u^2 + c).$$

So now the curvature formulas are verified. In fact, from the original parameterization (or via a simple calculation with Gauss and mean curvatures), we can show that the principal curvatures obey $k_\mu = 3\,k_\pi - 2c$, and this makes the rotating drop a linear Weingarten surface. For more information about these, see (Kuhnel and Steller 2005) and (Mladenov and Oprea 2007).

4.8.3 Parameterization via Legendre's Integrals

When performing the integration in (4.196), it turns out to be convenient to switch from variables $\tilde{a}, \tilde{c}$ to variables $\tau, \mathring{r}$ defined by the relations

$$\tilde{a}\mathring{r}^3 + \tilde{c}\mathring{r} = 1, \qquad \tau = \frac{1}{\tilde{a}\mathring{r}^3}. \tag{4.198}$$

In view of (4.194), the first equation has an obvious geometrical interpretation as specifying the distance $\mathring{r}$ from the origin to the point on the equator where the tangent is normal to the abscissa (cf. Fig. 4.22). The result of the integration is

$$z(r) = \frac{\mathring{r}^2 - \tilde{m} + \tilde{n}}{2\sqrt{\tilde{m}}} F(\varphi, k) + \sqrt{\tilde{m}}\left(E(\varphi, k) - \frac{\sin\varphi\sqrt{1 - k^2 \sin^2\varphi}}{1 + \cos\varphi}\right), \tag{4.199}$$

where

$$\tilde{m} = \sqrt{\tau^2 + 2\tau\,\mathring{r}^2}, \qquad \tilde{n} = (\tau - 1)\,\mathring{r}^2, \qquad k = \frac{1}{2}\sqrt{\frac{1 + 2\tau + 2\sqrt{\tau(\tau+2)}}{\sqrt{\tau(\tau+2)}}}$$

and

$$\varphi = \arccos\frac{\tilde{m} - \mathring{r}^2 + r^2}{\tilde{m} + \mathring{r}^2 - r^2}. \tag{4.200}$$

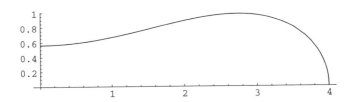

Fig. 4.22 A typical profile of the drop generated with $\mathring{r} = 4$ and $\tau = 0.533333$.

4.8.4 Parameterization via Weierstrass's Functions

The parameterization presented in the previous section is, however, not quite suitable for the main goal of this section of finding closed geodesics on the surface, since it can be used only for visualizing its upper part. To find geodesics, we need another parameterization which describes the entire surface, and that is exactly what we will do below.

Let us start by introducing new parameters, replacing $\tilde{a}$ and $\tilde{c}$ with a and c given by the formulas

$$\tilde{a} = \frac{a}{\mathring{r}^3}, \qquad \tilde{c} = \frac{c}{\mathring{r}}, \qquad a \in \mathbb{R}^+. \tag{4.201}$$

This choice is made because the geometrical relation (4.198) then reduces to

$$\theta(\mathring{r}) = \frac{\pi}{2}, \tag{4.202}$$

which also leads to a very simple relation between the new parameters, namely,

$$c = 1 - a. \tag{4.203}$$

Further on, it also turns out to be convenient to introduce the notation

$$\hat{r} = \frac{r}{\mathring{r}}, \qquad \hat{z} = \frac{z}{\mathring{r}}, \tag{4.204}$$

which allows us to rewrite (4.196) as

$$z = \mathring{r}\hat{z} = \mathring{r} \int \frac{(a\hat{r}^2 + 1 - a)\hat{r}d\hat{r}}{\sqrt{1 - \hat{r}^2(a\hat{r}^2 + 1 - a)^2}}, \tag{4.205}$$

and this also suggests the following change in the integral:

$$\chi = \hat{r}^2, \tag{4.206}$$

transforming it into the form

$$\hat{z} = \frac{1}{2} \frac{(a(\chi - 1) + 1)d\chi}{\sqrt{1 - \chi(a(\chi - 1) + 1)^2}}. \tag{4.207}$$

The expression under the radical is a polynomial of the third degree which could be easily converted into the Weierstrass' form by substituting the variable χ with ξ via the formula

$$\chi = m\xi + n, \tag{4.208}$$

where

$$m = -\left(\frac{2}{a}\right)^{2/3} \quad \text{and} \quad n = \frac{2(a-1)}{3a}. \tag{4.209}$$

In this manner, we arrive at an integrand of the type

$$\frac{d\xi}{\sqrt{4\xi^3 - g_2\xi - g_3}} = du, \tag{4.210}$$

where

$$g_2 = \frac{(a-1)^2}{3}\left(\frac{2}{a}\right)^{2/3}, \quad g_3 = \frac{(a+2)\left(2a^2 - 10a - 1\right)}{27a} \tag{4.211}$$

and u is some auxiliary (actually uniformizing) variable.

The integration of (4.210) is equivalent to inverting the Weierstrass integral of the first kind provided by the Weierstrass $\wp$ function, i.e.,

$$\xi = \wp(u, g_2, g_3). \tag{4.212}$$

By making further use of the algebra and the properties of the Weierstrass functions, we end up with the expressions (see also the generated by them profile curves in Fig. 4.23)

$$r(u) = \mathring{r}\sqrt{m\wp(u, g_2, g_3) + n} \tag{4.213}$$

$$z(u) = \mathring{r}\left(\frac{a-1}{6}\left(\frac{2}{a}\right)^{2/3} u - \left(\frac{2}{a}\right)^{1/3}\zeta(u, g_2, g_3)\right).$$

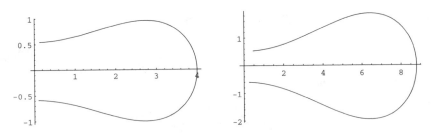

Fig. 4.23 Profiles of the drop generated via (4.213) with $\mathring{r} = 4$, $a = 1/0.53333$ (*Left*) and $\mathring{r} = 8.8$, $a = 1/0.47$ (*Right*)

4.8.5 Geodesics on the Drop

The full parameterization of the rotating drop given in the previous section allows us to visualize geodesics on the rotating drop. Similar visualizations and analyses have previously appeared in Alexander (2006), Borzellino et al. (2007), Mladenov and Oprea (2003) and Mladenov and Oprea (2003a) and (Oprea, 2007). The key point is that the Clairaut relation for a surface of revolution always confines geodesics that start parallel to parallels to lying between parallels! This allows for an algorithm to be developed that *finds closed geodesics*. This algorithm is called a "halfbouncepoint" and it relies on the symmetry and the Intermediate Value Theorem to find an initial condition for a closed geodesic. The geodesic is started parallel to a parallel at some u_0 and then travels halfway around the surface where u_1 is measured. By symmetry, if the geodesic is closed, then we must have $u_1 = u_0$. The computer algorithm (implemented on Maple) numerically solves for u_1 and then computes $u_0 - u_1$ for values of u surrounding u_0. If both positives and negatives occur, then the Intermediate Value Theorem says that there is a point where $u_1 = u_0$, and we have found a closed geodesic. The analyses in Alexander [2006] and Borzellino *et al.* [2007] validate the use of the Intermediate Value Theorem for surfaces of revolution. The rest of this section gives the Maple approach to finding closed geodesics on the drop. So let us now give procedures that plot geodesics on surfaces.

```
>  with(LinearAlgebra):with(plots):
>  geoeq:=proc(X)
>  local M,eq1,eq2;
>  M:=EFG(X);
>  Eq.1:=diff(u(t),t$2)+subs({u=u(t),v=v(t)},
>  diff(M[1],u)/(2*M[1]))*diff(u(t),t)^2 + subs({u=u(t),v=v(t)},
>  diff(M[1],v)/(M[1]))*diff(u(t),t)*diff(v(t),t)-subs({u=u(t),
>  v=v(t)}, diff(M[3],u)/(2*M[1]))*diff(v(t),t)^2=0;
>  Eq.2:=diff(v(t),t$2)-subs({u=u(t),v=v(t)},
>  diff(M[1],v)/(2*M[3]))*diff(u(t),t)^2+subs({u=u(t),v=v(t)},
>  diff(M[3],u)/(M[3]))*diff(u(t),t)*diff(v(t),t)+
>  subs({u=u(t),v=v(t)},diff(M[3],v)/(2*M[3]))*diff(v(t),t)^2=0;
>  Eq.1,eq2;
>  end:
>  plotgeo:=proc(X,ustart,uend,vstart,vend,u0,v0,Du0,Dv0,
>  T,N,gr,theta,phi)
>  local sys,desys,u1,v1,listp,geo,plotX;
>  sys:=geoeq(X);
>  desys:=dsolve({sys,u(0)=u0,v(0)=v0,D(u)(0)=Du0,D(v)(0)=Dv0},
>  {u(t),v(t)}, type=numeric, output=listprocedure);
>  u1:=subs(desys,u(t)); v1:=subs(desys,v(t));#####
>  print(u1(0.1),v1(0.1));
>  geo:=tubeplot(subs({u='u1'(t),v='v1'(t)},X),t=0..T,
>  radius=0.02,color=black,thickness=2,numpoints=N):
>  plotX:=plot3d(subs({u=u(t),v=v(t)},X),u=ustart..uend,
>  v=vstart..vend, grid=[gr[1],gr[2]]):
>  display({geo,plotX}, shading=XY,lightmodel=light2,
>  scaling=constrained,orientation=[theta,phi]);
>  end:
```

With the parameterization of the preceding section, we can visualize the profile curve and drop surface using Maple's Weierstrass functions.

```
>   surfHprofile:=proc(a,r0)
>   local m,n,g2,g3;
>   m:=-(2/a)^(2/3);
>   n:=2*(a-1)/(3*a);
>   g2:=(a-1)^2/3*(2/a)^(2/3);
>   g3:=(a+2)*(2*a^2-10*a-1)/(27*a);
>   umax:=fsolve(diff(r0*((a-1)/6*(2/a)^(2/3)*u
>   -(2/a)^(1/3)*WeierstrassZeta(u,g2,g3)),u)=0,u=0..4);
>   plot([r0*sqrt(m*WeierstrassP(u,g2,g3)+n),
>   r0*((a-1)/6*(2/a)^(2/3)*u
>   -(2/a)^(1/3)*WeierstrassZeta(u,g2,g3)),u=0..4],
>   scaling=constrained,grid=[40,40]);
>   end:
```

(In fact, to get nice pictures, we need to know where u ends. Above, the procedure surfHprofile has $u = 0.4$, but 4 may not be optimal. The following procedure will give the u-value needed to get to any distance r along the horizontal axis. If $\mathring{r}$ is desired, then this will be halfway to the final u needed for plotting in surfHprofile.

```
>   uvalue2:=proc(a,r0,r)
>   local m,n,g2,g3;
>   m:=-(2/a)^(2/3);
>   n:=2*(a-1)/(3*a);
>   g2:=(a-1)^2/3*(2/a)^(2/3);
>   g3:=(a+2)*(2*a^2-10*a-1)/(27*a);
>   fsolve(r0*sqrt(m*WeierstrassP(u,g2,g3)+n)=r,u,0..4);
>   end:
```

For instance, to plot surfHprofile(1/.47,8.8), we find uvalue2(1/.47,8.8,8.79) (where we must take 8.79 instead of 8.8 because of numerical solving issues). We get the following:

```
>   uvalue2(1/.47,8.8,8.79);
```

$$2.702546206$$

In surfHprofile, we now replace $u = 0..4$ with $u = 0..2 * 2.702546206$.) Here are two profile curves obtained in this way.

```
>   surfHprofile(1/.53333,4);
>   surfHprofile(1/.47,8.8);
```

In the following procedure, which plots the drop surface for given a and r_0, the parameter p allows for switching from a constrained view of the surface when $p = 0$ to an unconstrained view when $p = 1$.

```
>   surfH:=proc(a,r0,p)
>   local m,n,g2,g3;
>   m:=-(2/a)^(2/3);
>   n:=2*(a-1)/(3*a);
>   g2:=(a-1)^2/3*(2/a)^(2/3);
>   g3:=(a+2)*(2*a^2-10*a-1)/(27*a);
>   if p=0 then
>   plot3d(subs({r=r(t),v=v(t)},
>   [r0*sqrt(m*WeierstrassP(u,g2,g3)+n)*cos(v),
>   r0*sqrt(m*WeierstrassP(u,g2,g3)+n)*sin(v),
>   r0*((a-1)/6*(2/a)^(2/3)*u
>   -(2/a)^(1/3)*WeierstrassZeta(u,g2,g3))]),u=0..4,v=0..2*Pi,
>   grid=[40,25],scaling=constrained,axes=boxed,
>   orientation=[58,74]);else
>   plot3d(subs({u=u(t),v=v(t)},
>   [r0*sqrt(m*WeierstrassP(u,g2,g3)+n)*cos(v),
>   r0*sqrt(m*WeierstrassP(u,g2,g3)+n)*sin(v),
>   r0*((a-1)/6*(2/a)^(2/3)*u
>   -(2/a)^(1/3)*WeierstrassZeta(u,g2,g3))]),u=0..4,v=0..2*Pi,
>   grid=[40,25],scaling=unconstrained,axes=boxed,
>   orientation=[58,74]);
>   fi;
>   end:
```

Here are the rotating drop surfaces corresponding to the profile curves depicted above.

```
>   surfH(1/0.5333,4,0); surfH(1/0.47,8.8,0);
```

We now give a procedure that creates the total parameterization for a drop with given a and $\mathring{r}$.

```
>   rotdrop2:=proc(a,r0)
>   local m,n,g2,g3;
>   m:=-(2/a)^(2/3);
>   n:=2*(a-1)/(3*a);
>   g2:=(a-1)^2/3*(2/a)^(2/3);
>   g3:=(a+2)*(2*a^2-10*a-1)/(27*a);
>   [r0*sqrt(m*WeierstrassP(u,g2,g3)+n)*cos(v),
>   r0*sqrt(m*WeierstrassP(u,g2,g3)+n)*sin(v),
>   r0*((a-1)/6*(2/a)^(2/3)*u
>   -(2/a)^(1/3)*WeierstrassZeta(u,g2,g3))];
>   end:
```

The following geodesic plot shows that starting inside the drop hump produces wild geodesics (see Fig. 4.24).

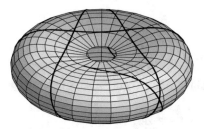

Fig. 4.24 A wild geodesic on the drop given by (4.213) with $\mathring{r} = 4$ and $a = 1/0.5333$

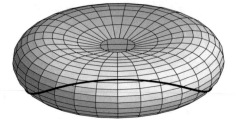 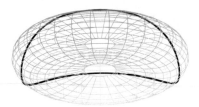

Fig. 4.25 A closed geodesic on a drop generated via (4.213) with $\mathring{r} = 4$ and $a = 1/0.5333$

```
>    plotgeo(rotdrop2(1/.5333,4),0,4,0,2*Pi,1.65,0,0,1,40,200,
>    [35,30],31,60);
```

Now let us give the procedure that will allow us to detect closed geodesics (see Mladenov and Oprea 2003a and Oprea 2007 and Fig. 4.25).

```
>    halfbouncepoint:=proc(u0,a,r0)
>    local geosystem,uuu,vvv,ttt,uuu0;
>    geosystem:=dsolve({geoeq(rotdrop2(a,r0)),u(0)=u0,v(0)=0,
>    D(u)(0)=0,D(v)(0)=1},{u(t),v(t)},
>    type=numeric, output=listprocedure,range=0..25);
>    uuu:=subs(geosystem,u(t)); vvv:=subs(geosystem,v(t));
>    ttt:=timelimit(30,fsolve(vvv(t)=evalf(Pi),t));
>    if type(ttt,float) then
>    uuu0:=eval(uuu(t),t=ttt);
>    evalf(u0-uuu0);
>    else print('time error');
>    end if;
>    end:
```

Here is how we use the procedure. After some preliminary guesses (and plots), we guess that a closed geodesic might be near $u_0 = 2.3$. We now use "halfbouncepoint" to calculate $u_0 - u_1$ as described earlier.

```
>    for jj from -2 to 2 do
>    print(2.3+jj*0.005,halfbouncepoint(2.3+jj*0.005,
>    1/.53333,4));
>    od;
```

$$2.290, \ -0.000011966$$
$$2.295, \ 0.602 \ 10^{-6}$$
$$2.3, \ -0.000040570$$
$$2.305, \ -0.000132853$$
$$2.310, \ -0.000273612$$

But now, the Intermediate Value Theorem says that we should look around $u_0 = 2.295$. We zoom in.

```
>    for jj from -2 to 2 do
>    print(2.292+jj*0.0005,halfbouncepoint(2.292+jj*0.0005,
>    1/.53333,4));
>    od;
```

$$2.2910, -0.5027\ 10^{-5}$$

$$2.2915, -0.2397\ 10^{-5}$$

$$2.292, -0.322\ 10^{-6}$$

$$2.2925, 0.1199\ 10^{-5}$$

$$2.2930, 0.2170\ 10^{-5}$$

So now we try near 2.2925, say 2.2923, and we see that we have an approximation to a closed geodesic.

```
>   plotgeo(rotdrop2(1/.5333,4),0,4,0,2*Pi,2.2923,0,0,1,28,200,
>   [35,30],0,67);
>   plotgeo(rotdrop2(1/.5333,4),0,4,0,2*Pi,2.2923,0,0,1,28,200,
>   [35,30],0,67);
```

But we saw above that there were actually two changes of sign in halfbouncepoint. So there is another closed geodesic to be found.

```
>   for jj from -2 to 2 do
>   print(2.2951+jj*0.0001,halfbouncepoint(2.2951+jj*0.0001,
>   1/.53333,4));
>   od;
```

$$2.2949, 0.886\ 10^{-6}$$

$$2.2950, 0.602\ 10^{-6}$$

$$2.2951, 0.296\ 10^{-6}$$

$$2.2952, -0.31\ 10^{-7}$$

$$2.2953, -0.379\ 10^{-6}$$

```
>   plotgeo(rotdrop2(1/.5333,4),0,4,0,2*Pi,2.29515,0,0,1,28,200,
>   [35,30],0,67);
>   plotgeo(rotdrop2(1/.5333,4),0,4,0,2*Pi,2.29515,0,0,1,28,200,
>   [35,30],0,67);
```

4.8.6 Questions for Future Work

The results above leave some questions unanswered. For instance, we seem to have a fundamental difference in shape for surfaces (or, indeed, their profile curves) when the parameter a obeys $a < 1$ and $a \geq 1$. For $a < 1$, the surfaces are prolate (i.e., the profiles do not have maxima). Yet, from the formulas, it is difficult to see why this is true. Even the formulas for Gauss and mean curvatures are simple only for the original parameterization and not for the parameterization using Weierstrass functions. Consider the following examples.

Here, we show the pictures for $a = 1.4$, because they are easier to see than values of a closer to 1. These elicit the following.

Question 4.1 *What is the explanation for the bifurcation at $a = 1$ (sphere) between drop surface shapes?*

In principle, we should be able to take $a \in (0, 4)$, but we get self-intersecting profile curves when a reaches the critical value $a = 2.32911$. This value was found using the `FindRoot` routine of `Mathematica`. For this value of a, the surface becomes internally tangent to itself, and for larger values, we get self-intersecting profile curves and their corresponding surfaces.

Question 4.2 *What is the explanation for the bifurcation around* $a = 2.32911$ *between profile shapes? Furthermore, what is the exact bifurcation parameter* a *from a theoretical (i.e., analytical rather than numerical) viewpoint?*

Finally, in regard to geodesics, we see another difference in behavior between the cases $a \leq 1$ and $a > 1$. As we have seen, for $a > 1$, we can show the existence of non-equatorial closed geodesics and then "find" them using numerical solutions of the geodesic equations and fundamental theorems such as the Intermediate Value Theorem. For $a < 1$, we have been unsuccessful in finding any closed geodesic other than the equator. Of course, our approach to finding closed geodesics only looks for ones that start parallel to parallels, so perhaps there are others.

Question 4.3 *Are there closed non-equatorial geodesics on drop surfaces with* $a < 1$?

Perhaps the methods of (Alexander, 2006) may say something here by analyzing the Weierstrass functions in this context. The Clairaut relation in this context says that, if we start at $u = \bar{u}$ and parallel to a parallel, then along the geodesic, it must always be true that $G(u) \geq G(\bar{u})$, where $G(u)$ is the metric coefficient G at parameter value u. (Recall that a surface of revolution with revolution parameter v and profile curve parameter u has the property that the metric coefficients E and G only depend on u and the metric coefficient $F = 0$.) The G here (which is obtained from a bit of simplification applied to the result of the EFG procedure introduced earlier) is

$$- \left((2/a)^{2/3} . \wp (u, g_2, g_3) - \frac{2a - 2}{3a} \right) \mathring{r}^2, \tag{4.214}$$

which can be easily recognized as $r^2(u)$ from (4.213) and is in complete agreement with the expression given in (4.197). This means that a geodesic starting parallel to a parallel will always be confined to the region with $r(u) \geq r(\bar{u})$, that is, the geodesic stays at least as far from the axis of revolution as the starting point is.

The more surprising result produced by Maple is that E takes the form

$$\frac{1}{(2a^2)^{2/3}} \frac{\mathring{r}^4}{r^2(u)}. \tag{4.215}$$

But now we have the result.

Theorem 4.2 *The surface area* $\mathcal{A}$ *of the rotating drop with parameters* a *and* $\mathring{r}$ *is*

$$\mathcal{A} = 2\pi \left(\frac{2}{a}\right)^{2/3} \mathring{u}\,\mathring{r}^2, \tag{4.216}$$

where $\mathring{u}$ satisfies $r(\mathring{u}) = \mathring{r}$.

Proof Since the metric coefficient F vanishes for a surface of revolution, the surface area is given by

$$2\int_0^{2\pi}\int_0^{\mathring{u}} \sqrt{E\,G}\,du\,dv, \tag{4.217}$$

where the factor 2 accounts for the symmetric bottom half of the drop. But by the above, we see that $\sqrt{E\,G} = \mathring{r}^2/(2a^2)^{1/3}$, a constant. The integration is then trivial.

The fact that the total surface area of the drop is proportional to $\mathring{u}$ is quite unusual and interesting. It raises the question:

Question 4.4 *Does the formula for surface area have any physical consequences? Find other surfaces with this relationship. Does the proportionality of surface area to $\mathring{u}$ characterize a class of surfaces?*

We have seen that physical systems produce interesting geometry and that this geometry can be studied if we can parameterize properly – even if we must use complicated functions such as Weierstrass elliptic functions. We have also seen that computers and modern computer algebra software can provide insights through visualization that might have eluded us in the past.

References

J. Alexander, Closed geodesics on certain surfaces of revolution. J. Geom. Symmetry Phys. **8**, 1–16 (2006)

F. Baginski, On the design and analysis of inflated membranes: natural and pumpkin shaped balloons. SIAM J. Appl. Math. **65**, 838–857 (2005)

F. Baginski, J. Winker, The natural shape balloon and related models. Adv. Space Res. **65**, 1617–1622 (2004)

J. Bernoulli, Quadratura curvae, e cujus evolutione describitur inflexae laminae curvatura. Birkhäuser 223–227 (1694)

G. Birkhoff, C. de Boor, Piecewise polynomial interpolation and approximation. in ed. By H. Garbedian *Approximation of Functions, Proc. General Motors Symposium of 1964*, (Elsevier, Amsterdam, 1965), pp. 164–190

M. Born, Untersuchungen über die Stabilität der elastischen Linie in Ebene und Raum, unter Verschiedenen Grenzbedingungen. Ph.D. thesis, University of Göttingen (1906)

G. Borzellino, C. Jordan-Squire, D. Sullivan, Closed geodesics on orbifolds of revolution. Houston J. Math. **33**, 1011–1025 (2007)

P. Byrd, M. Friedman, Handbook of Elliptic Integrals for Engineers and Scientists (Springer, Berlin, 1954)

B. Cox, J. Hill, A variational approach to the perpendicular joining of nanotubes to plane sheets. J. Phys. A Math. Theor. **41**(125203), 1–11 (2008)

L. d'Antonio, The fabric of the universe is most perfect: euler's research on elastic curves. in Euler at 300 (Mathematical Association of America, Washington, 2007), pp. 239–260

C. Delaunay, Sur la surface de revolution dont la courbure moyenne est constante. J. Math. Pures et Appliquées **6**, 309–320 (1841)

P. Djondjorov, M. Hadzhilazova, I. Mladenov, V. Vassilev, Explicit parameterization of Eulers elastica. Geom. Integr. Quant. **9**, 175–186 (2008)

P. Djondjorov, M. Hadzhilazova, I. Mladenov, V. Vassilev, A note on the passage from the free to the elastica with a tension. Geom. Integr. Quant. **10**, 175–182 (2009)

J. Eells, The surfaces of Delaunay. Math. Intell. **9**, 53–57 (1987)

L. Euler, Methodus inveniendi lineas curvas maximi minimive proprietate gaudentes, sive solutio problematis isoperimetrici lattissimo sensu accepti. Additamentum (1744)

G. Gibbons, The shape of a mylar balloon as an elastica of revolution. (Cambridge University, Cambridge 7pp, 2006). DAMTP Preprint

V. Goss, The history of the planar elastica: Insights into mechanics and scientific method. Sci. Educ. **18**, 1057–1082 (2009)

I. Gradshteyn, I. Ryzhik, *Table of Integrals, Series and Products*, 7th edn. (Academic Press, New York, 2007)

G. Grason, C. Santangelo, Undulated cylinders of charged diblock copolymers. Eur. Phys. J. E **20**, 335–346 (2006)

A. Gray, *Modern Differential Geometry of Curves and Surfaces with Mathematica*, 2nd edn. (CRC Press, Boca Raton, 1998)

M. Hadzhilazova, I. Mladenov, On evolutes of nodary and undulary Delaunay curves. in *Proceedings of the Thirty Eighth Spring Conference of the Union of Bulgarian Mathematicians* (UBM, Sofia, 2009), pp. 131–137

M. Hadzhilazova, I. Mladenov, Once more the mylar balloon. CRAS (Sofia) **61**, 847–856 (2008)

M. Hadzhilazova, I. Mladenov, Membrane approach to balloons and some related surfaces. Geom. Integr. Quant. **7**, 176–186 (2006)

M. Hadzhilazova, I. Mladenov, J. Oprea, Unduloids and their geometry. Arch. Math. **43**, 417–429 (2007)

H. Irvine, *Cable Structures, MIT Series in Structural Mechanics* (MIT Press, MA, 1981)

E. Jahnke, F. Emde, F. Lösch, *Tafeln Höherer Funktionen* (Teubner, Stuttgart, 1960)

M. Koiso, B. Palmer, Equilibria for anisotropic energies and the Gielis formula. Forma **23**, 1–8 (2008)

W. Kuhnel, M. Steller, On closed Weingarten surfaces. Monatsh. Math. **146**, 113–126 (2005)

R. Lopez, On linear Weingarten surfaces. Int. J. Math **19**, 439–448 (2008)

A.E.H. Love, *A Treatise on the Mathematical Theory of Elasticity* (Dover, New York, 1944)

A. Ludu, C. Cibert, Elastic axonemal model: Solitary wave shapes. Math. Comp. Sim. **80**, 223–230 (2009)

P. Marinov, M. Hadzhilazova, I. Mladenov, Elastic Sturmian spirals. Compt. Rend. de l'Acad. Bulg. des Sci. **67**, 167–172 (2014)

S. Matsutani, Euler's elastica and beyond. J. Geom. Symmetry Phys. **17**, 45–86 (2010)

I. Mladenov, Delaunay surfaces revisited. CRAS (Sofia) **55**, 19–24 (2002a)

I. Mladenov, M. Hadzhilazova, P. Djondjorov, V. Vassilev, On the intrinsic equation behind the Delaunay surfaces. in *AIP Conference Proceedings*, vol. 1079

I. Mladenov, J. Oprea, Unduloids and their closed geodesics. Geom. Integr. Quanti. **4**, 206–234 (2003a)

I. Mladenov, J. Oprea, The mylar balloon revisited. Am. Math. Month. **110**, 761–784 (2003)

I. Mladenov, J. Oprea, On some deformations of the mylar balloon. Publ. de la RSME **11**, 310–315 (2007)

I. Mladenov, M. Hadzhilazova, P. Djondjorov, V. Vassilev, On the generalized Sturmian spirals. C. R. Bulg. Acad. Sci. **64**, 633–640 (2011a)

J. Oprea, *Differential Geometry and Its Applications*, 3rd edn. (Mathematical Association of America, Washington D.C., 2007)

J. Oprea, *The Mathematics of Soap Films: Explorations with Maple* (AMS, Providence, 2000)

W. Paulsen, What is the shape of the mylar balloon? Amer. Math. Month. **101**, 953–958 (1994)

E. Popova, M. Hadzhilazova, I. Mladenov, On balloons, membranes and surfaces representing them. in *Proceedings of the Thirty Fifth Spring Conference of the Union of Bulgarian Mathematicians* (UBM, Sofia, 2006), pp. 158–164

D. Singer, Lectures on elastic curves and rods. AIP Conf. Proc. **1002**, 3–32 (2008)

V. Vassilev, P. Djonjorov, I. Mladenov, Cylindrical equilibrium shapes of fluid membranes. J. Phys. A Math. Theor. **41**, 435201–16pp (2008)

W. Whewell, On the intrinsic equation of a curve, and its application. Camb. Phil. Soc. **8**, 659–671 (1849)

W. Whewell, Second memoir on the intrinsic equation of a curve and its application. Camb. Phil. Trans. **9**, 150–156 (1850)

Chapter 5
Equations of Equilibrium States of Membranes

Abstract The energy of the bending of the membranes within the Canham model is directly justified as an extension of the case of bending of the elastic strips known as Euler's Elasticas. Later, this model was elaborated upon by Helfrich & Deuling in the form of a non-linear system of two equations for the curvatures of the axially-symmetric membranes. Finally, Ou-Yang and Helfrich introduced the model which is currently seen to be most adequate for the description of membrane configurations. Since we are primarily interested in analytical solutions, it is quite natural that the equation by Ou-Yang and Helfrich be examined for the presence of symmetries, because they are in direct connection with solutions. When this is done, it becomes clear that the most general group of symmetries of the shape equation of the form coincides with the group of Euclidean motions in the real three-dimensional space. Among the generators of this group are those of rotations and translations, which hint about the existence of the analytical solutions discussed in the present and the subsequent chapters.

5.1 Canham Model

The evaluation of the stored elastic energy in a thin plate bent into two perpendicular planes is very complicated. However, if the shear stresses are zero, it would appear that the problem reduces to the integral of the sum of the squares of the two curvatures (Fig. 5.1).

It has been necessary to deviate considerably from the equations of structural engineering in order to accommodate the abundant biological observations, e.g., the red blood cells (RBC) can adopt so many shapes without hemolysis and can change from the crenated form into a biconcave shape.

To solve the problem concretely with an RBC shape, Canham (1970) had considered the membrane's elastic energy of bending in the form

$$U = \frac{D}{2} \int (\kappa_1^2 + \kappa_2^2) \mathrm{d}\mathcal{A}, \tag{5.1}$$

where D is the bending rigidity, and κ_1 and κ_2 are the principle curvatures of the surface based on the model described in the next section.

© Springer International Publishing AG 2017
I.M. Mladenov and M. Hadzhilazova, *The Many Faces of Elastica*,
Forum for Interdisciplinary Mathematics 3, DOI 10.1007/978-3-319-61244-7_5

Fig. 5.1 A section of a
membrane which is curved
in two planes. The two
curvature centers must lie on
the normal to the surface

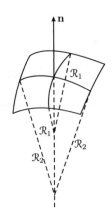

5.2   Key Assumptions in the Model

(1) The membrane consists mainly of two isotropic labile surfaces - an interpretation
supported by the electron micrographs of membranes and the viscoelastic studies by Rand (1964). It is acknowledged that micellar structures also exist in the
membrane, but since they are less stable, Davson and Danielli (1952) assumed
that the greater part of the membrane surface has the bimolecular leaflet arrange-
ment.

The interpretation of the viscoelastic nature of the red-cell membrane is illustrated in Fig. 5.2 (one might consider the stress state in the membrane as a
hydrostatic in two dimensions).

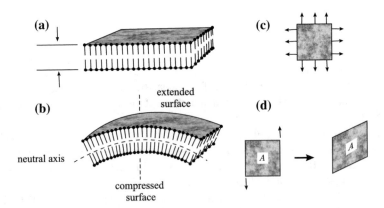

Fig. 5.2 A schematic interpretation of the RBC membrane. **a** the leaflet model proposed by Davison
and Danielli. **b** An element of membrane has been bent in one plane, showing the compression of
the inner surface and the stretching of the outer surface. **c** An element of the area which is stressed
equally in all directions in the plane of the membrane (a two-dimensional hydrostatic stress) will
resist the stress elastically. **d** The membrane will offer only viscous resistance to shear stresses,
resulting in deformation with no storage of elastic energy

The flat membrane can resist distortion only temporarily, without area changes, in the form of a viscous resistance, but can resist changes in, the area. Any attempt to change the area of either membrane side leads to an elastic resistance to bending, because the inner surface is compressed and the outer surface is extended (Fig. 5.2).

(2) It is not assumed that the red cell is cast in the biconcave shape, but rather that the biconcave shape minimizes the elastic energy stored in the membrane. We assumed that an element of the membrane has no stored bending energy if it is flat, and that any curved element of the membrane has stored elastic energy.

(3) Any change of the shape due to the osmotic volume shifts are considered to take place without altering the total cellular area (that is, the area of the surface through the neutral axis (Fig. 5.2).

(4) The membrane has the same physical properties over the entire surface. This concurs with the results of Rand and Burton (1964), but contradicts those of Murphy (1965) and Fung and Tong (1968).

5.3 Helfrich and Deuling Model

To describe the shape of RBC, Helfrich and Deuling (1975) suggest a generalization of Cahnam's model.

In this model, the energy density per unit area has the form

$$\mathcal{E} = \frac{1}{2}k_c(\kappa_1 + \kappa_2 - h)^2 + \frac{1}{2}\bar{k}_c\kappa_1\kappa_2, \tag{5.2}$$

where κ_1 and κ_2 are the principal curvatures, k_c and $\bar{k}_c$ are the membrane's elastic moduli and h is the so-called spontaneous curvature.

For closed surfaces (like RBC), the integral of the second term will be a constant, and consequently, the shape of the membrane is determined only by the first term. Taking into account the axial symmetry of RBC, we can write

$$\delta\left(\frac{1}{2}k_c\int(\kappa_\mu + \kappa_\pi - h)^2 d\mathcal{A} + \Delta pV + \lambda\mathcal{A}\right) = 0. \tag{5.3}$$

Here, κ_μ and κ_π are, respectively, the curvatures along the meridians and the parallels (see (1.80)), and Δp and λ are the Lagrange multipliers. If we choose x (the distance from the axis of symmetry OZ) as a parameter, the formulas for the curvatures are

$$\kappa_\mu = \frac{\ddot{z}(x)}{(1 + \dot{z}^2(x))^{3/2}}, \qquad \kappa_\pi = \frac{\dot{z}(x)}{x\sqrt{1 + \dot{z}^2(x)}}. \tag{5.4}$$

Besides, we have the identity which can be checked directly, i.e.,

Fig. 5.3 Profile curve and
geometry of axially
symmetric surface

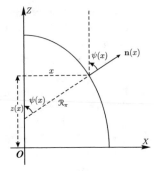

$$\kappa_\mu = \frac{d}{dx}(x\kappa_\pi), \tag{5.5}$$

and the expanded form of which is

$$\frac{d\kappa_\pi}{dx} = \frac{\kappa_\mu - \kappa_\pi}{x}. \tag{5.6}$$

If we denote by $\psi(x)$ the angle made by the surface normal and the axis of rotation and consider the triangle in which the hypotenuse is the curvature radius $\mathcal{R}_\pi$, we can write the equation (cf. Fig. 5.3)

$$\frac{x}{\mathcal{R}_\pi} = \sin\psi, \tag{5.7}$$

and this relation can be immediately transformed into the form

$$\kappa_\pi = \frac{\sin\psi}{x}. \tag{5.8}$$

Combining this result with (5.5), we have the expression for the other curvature as well, i.e.,

$$\kappa_\mu = \cos\psi \frac{d\psi}{dx}. \tag{5.9}$$

These equations should be supplemented with the geometrical relation

$$\frac{dz}{dx} = -\tan\psi(x), \tag{5.10}$$

in which the sign accounts for the fact that when x is increasing, z is decreasing. Respectively, the volume V and the surface area $\mathcal{A}$ are defined by the integrals

$$V = \int dV \quad \text{and} \quad \mathcal{A} = \int d\mathcal{A}, \tag{5.11}$$

in which the formulas for the infinitesimal quantities of the volume and the area are

$$dV = \pi x^2 |dz| = \frac{\pi x^3 \kappa_\pi dx}{\sqrt{1 - x^2 \kappa_\pi^2}}, \quad d\mathcal{A} = 2\pi x ds = \frac{2\pi x dx}{\sqrt{1 - x^2 \kappa_\pi^2}}. \tag{5.12}$$

Using the relation (5.6), we can eliminate κ_μ in Eq. (5.3) and in this way, transform it into the form

$$\delta \int \frac{x((x\dot{\kappa}_\pi + 2\kappa_\pi - \hbar)^2 + \kappa_\pi + \Delta p x^2 \kappa_\pi / k_c + 2\lambda/k_c)}{\sqrt{1 - x^2 \kappa_\pi^2}} dx = 0, \tag{5.13}$$

where $\dot{\kappa}_\pi = \dfrac{d\kappa_\pi}{dx}$.

The Euler-Lagrange equation corresponding to the integrand in (5.13), which is already in the canonical form $F = F(x, \kappa_\pi, \dot{\kappa}_\pi)$, can be written as

$$\dot{\kappa}_\mu = \frac{d\kappa_\mu}{dx} = \frac{x((\kappa_\pi(\kappa_\pi - \hbar)^2 - \kappa_\mu^2)/2 + (\lambda/k_c)\kappa_\pi + (1/2)\Delta p/k_c)}{\sqrt{1 - x^2 \kappa_\pi^2}}$$
$$- \frac{\kappa_\mu - \kappa_\pi}{x}. \tag{5.14}$$

This equation is of the first order, because the second derivative of κ_π is expressed via $\dot{\kappa}_\mu$ by making use (twice) of the fundamental geometrical relation (5.6). Taken together, Eqs. (5.6) and (5.14) form a close system of ODE which can be integrated (numerically). One can find the membrane contour after one more integration, namely,

$$z(x) = -\int \tan \psi(x) dx = -\int \frac{x \kappa_\pi}{\sqrt{1 - x^2 \kappa_\pi^2}} dx. \tag{5.15}$$

This model was not developed further on because it is purely numerical, and after a few years, it was replaced by the Ou-Yang and Helfrich model, which will be considered in the next section.

5.4   Ou-Yang and Helfrich Model

The equilibrium shapes of a vesicle in this model (Helfrich 1973) are determined by
the minimization of the shape energy, which may be written as

$$F = \frac{1}{2}k_c \oint (\kappa_1 + \kappa_2 - \text{lh})^2 \mathrm{d}A + \Delta p \int \mathrm{d}V + \lambda \oint \mathrm{d}A. \tag{5.16}$$

Here (as before), $\mathrm{d}A$ and $\mathrm{d}V$ are the infinitesimal elements surface of, respectively,
the area and the volume, k_c is the bending rigidity, κ_1 and κ_2 are the two principal
curvatures, and lh is the spontaneous curvature. The latter serves to describe the effect
of the asymmetry of the membrane or its environment (Deuling and Helfrich, 1976).
The first term in (5.16) is the elastic energy of the vesicle. The second and third
terms take into account the constraints of constancy of, respectively, the volume
and the area or represent the actual work of the deformation. Depending on the
situation, the pressure difference $\Delta p = p_{\text{out}} - p_{\text{in}}$ and the tensile stress λ serve as
Lagrange multipliers or they are prescribed experimentally by measuring the volume
or the area. Instead of the last term of Eq. (5.16), Jenkins 1977a, b introduces a local
area constraint by $\gamma \oint \mathrm{d}A$, where γ is a Lagrangian multiplier which varies with the
position. He had derived a general equilibrium equation, but did not consider the
spontaneous curvature except in the special case of the fluctuating sphere (Schneider
et al. 1984). A generalized equilibrium shape equation including the spontaneous
curvature will be derived below.

5.4.1   Basic Formulas and Definitions

Theoretically, the membrane of any vesicle may be represented as a closed surface
in the Euclidean three space given by the vector $\mathbf{x}(u, v)$, depending on the two real
parameters u, v. We introduce the following quantities which will be used further
(see Chap. 1):

$$
\begin{aligned}
\mathbf{x}_i &= \partial_i \mathbf{x}, & \mathbf{x}_{ij} &= \partial_i \partial_j \mathbf{x}, & g_{ij} &= \mathbf{x}_i . \mathbf{x}_j \\
g^{ij} &= (g_{ij})^{-1}, & g &= \det(g_{ij}), & h_{ij} &= \mathbf{x}_{ij} . \mathbf{n} \\
h^{ij} &= (h_{ij})^{-1}, & h &= \det(h_{ij}), & i, j &= u, v = 1, 2,
\end{aligned}
\tag{5.17}
$$

where $\partial_1 = \partial_u$, $\partial_2 = \partial_v$, and the matrices g_{ij} and h_{ij} are associated with the first
and second fundamental forms of the surface. The outward unit normal vector $\mathbf{n}$,
and the Christoffel symbols Γ_{ij}^k are defined by (cf. (1.71))

$$\mathbf{n} = (\mathbf{x}_1 \times \mathbf{x}_2)/\sqrt{g}, \qquad \mathbf{x}_{ij} = \Gamma_{ij}^k \mathbf{x}_k + h_{ij}\mathbf{n}. \tag{5.18}$$

Here and immediately following the repeated indices imply summation over them. The mean and Gaussian curvatures can be written, respectively, as

$$H = \frac{1}{2}(\kappa_1 + \kappa_2) = \frac{1}{2}g^{ij}h_{ij}, \qquad K = \kappa_1\kappa_2 = h/g. \tag{5.19}$$

We assume $\mathbf{x}$ to be an equilibrium shape and consider a slightly distorted surface defined by the equations

$$\tilde{\mathbf{x}} = \mathbf{x} + \psi(u, v)\mathbf{n}, \tag{5.20}$$

where $\psi(u, v)$ is a sufficiently smooth function. Using Eqs. (5.17)–(5.20), we may calculate $\tilde{\mathbf{x}}_i, \tilde{\mathbf{x}}_{ij}, \tilde{g}_{ij}, \tilde{h}_{ij}$, and so on. For example, we have

$$\tilde{\mathbf{x}}_i = \mathbf{x}_i + \psi_i\mathbf{n} + \psi\partial_i\mathbf{n}, \tag{5.21}$$

where $\psi_i = \partial_i\psi$. By using the Weingarten equations (cf. (1.69))

$$\partial_i\mathbf{n} = -h_{ij}g^{jk}\mathbf{x}_k, \tag{5.22}$$

we can transform the expression at the far right-hand side of (5.21) into

$$\tilde{\mathbf{x}}_i = \mathbf{x}_i + \psi_i\mathbf{n} - \psi h_{ij}g^{jk}\mathbf{x}_k. \tag{5.23}$$

Further on, by using the identities

$$h_{ij}g^{jk}h_{kl} = 2Hh_{il} - Kg_{il} \tag{5.24}$$

and $\mathbf{x}_i.\mathbf{n} = 0$ allows us to write

$$\delta g_{ij} = \tilde{\mathbf{x}}_i.\tilde{\mathbf{x}}_j - \mathbf{x}_i.\mathbf{x}_j = -2\psi h_{ij} + \psi_i\psi_j + \psi^2(2Hh_{ij} - Kg_{ij}). \tag{5.25}$$

The identity

$$g = \det(g_{ij}) = \frac{1}{2}\varepsilon_{3ij}\varepsilon_{3kl}g_{ik}g_{jl} \tag{5.26}$$

in combination with (5.25) results in

$$\delta g = g(-4\psi H + g^{ij}\psi_i\psi_j + \psi^2(4H^2 + 2K)) + \mathcal{O}(\psi^3). \tag{5.27}$$

Here and immediately following $\mathcal{O}(\psi^3)$ refers to terms of higher than quadratic order in ψ, and the symbols e_{ijk} are defined as

$$e_{ijk} = \begin{cases} +1, & \text{when } ijk \text{ is an even permutation of 123} \\ -1, & \text{when } ijk \text{ is an odd permutation of 123} \\ 0, & \text{otherwise.} \end{cases} \tag{5.28}$$

In addition, we have the variations

$$\delta g^{ij} = 2\psi(2Hg^{ij} - Kh^{ij}) + \left(\frac{1}{g}\varepsilon_{3ik}\varepsilon_{3jl} - g^{ij}g^{kl}\right)\psi_k\psi_l$$
$$-3\psi^2\left((K - 4H^2)g^{ij} + 2HKh^{ij}\right) + \mathcal{O}(\psi^3) \tag{5.29}$$

$$\delta h_{ij} = \psi_{ij} + \psi(Kg_{ij} - 2Hh_{ij}) - \Gamma^k_{ij}\psi_k$$
$$+ \psi\psi_m\left(K\Gamma^k_{ij}\varepsilon_{3lk}\varepsilon_{3mp}g_{pq}h^{ql} + (h_{ij}g^{lm})_l + h_{jk}g^{kl}\Gamma^m_{li}\right) \tag{5.30}$$
$$+ \psi_k\psi_m\left(g^{lk}(\delta_{im}h_{jl} + \delta_{jm}h_{il} - \frac{1}{2}h_{ij}g^{mk}\right) + \mathcal{O}(\psi^3).$$

Taking into account the equations $(h_{jl}g^{lm})_i = \partial_i(h_{jl}g^{lm})$, the definition of the Kronecker symbol

$$\delta_{ij} = \begin{cases} 0, & i \neq j \\ 1, & i = j \end{cases}$$

and Eqs. (5.19), (5.29) and (5.30), one can obtain the variation of the mean curvature

$$\delta H = \psi(2H^2 - K) + \frac{1}{2}g^{ij}\nabla_i\psi_j + \psi^2(4H^3 - 3HK)$$
$$+ \frac{1}{2}\psi\psi_m\left(g^{ij}(h_{jl}g^{lm})_i - h_{lk}g^{km}g_{ij}\Gamma^h_{ij} + (Kh^{ij} - 2Hg^{ij})\Gamma^m_{ij}\right) \tag{5.31}$$
$$+ \frac{1}{2}\psi_i\psi_j(Hg^{ij} - Kh^{ij}) + \psi\psi_{ij}(2Hg^{ij} - Kh^{ij}) + \mathcal{O}(\psi^3),$$

where $\nabla_i\psi_j$ is the covariant derivative of ψ_j defined by

$$\nabla_i\psi_j = \psi_{ij} - \Gamma^k_{ij}\psi_k. \tag{5.32}$$

The local variation of the surface area is given by

$$\delta\sqrt{g} = (-2\psi H + \frac{1}{2}g^{ij}\psi_i\psi_j + \psi^2 K)\sqrt{g} + \mathcal{O}(\psi^3), \tag{5.33}$$

and the global one by the formula

$$\delta A = \delta\oint \mathrm{d}A = \oint(-2\psi H + \frac{1}{2}g^{ij}\psi_i\psi_j + \psi^2 K)\mathrm{d}A + \mathcal{O}(\psi^3). \tag{5.34}$$

The variation of the volume is found to be

$$\delta V = \oint(\psi - \psi^2 H)\mathrm{d}A + \mathcal{O}(\psi^3), \tag{5.35}$$

and evidently all variations that have just been derived can be expressed via H, K, g_{ij}, h_{ij} and Γ_{ij}.

5.4.2 Shape Equation

In order to obtain the equation of mechanical equlibrium of the vesical membrane, we have to calculate the first variation of the shape energy given by Eq. (5.16). Because of (5.19), we may write

$$\delta F = \frac{1}{2}k_c\delta\oint(2H+\text{Ih})^2\mathrm{d}\mathcal{A} + \Delta p\,\delta\int\mathrm{d}V + \lambda\delta\oint\mathrm{d}\mathcal{A}. \qquad (5.36)$$

The first variations of V and $\mathcal{A}$ are defined in (5.34) and (5.35) and, respectively,

$$\delta\oint\mathrm{d}\mathcal{A} = -\oint 2\psi H\mathrm{d}\mathcal{A} \qquad (5.37)$$

and

$$\delta\int\mathrm{d}V = \oint\psi\mathrm{d}\mathcal{A}. \qquad (5.38)$$

The first variation of the curvature-elastic energy may be written as

$$\begin{aligned}\delta F_c &= \frac{1}{2}k_c\delta\oint(2H+\text{Ih})^2\mathrm{d}\mathcal{A} \\ &= \frac{1}{2}k_c\oint\left((2H+\text{Ih})^2\delta\mathrm{d}\mathcal{A} + 4(2H+\text{Ih})\delta H\,\mathrm{d}\mathcal{A}\right). \qquad (5.39)\end{aligned}$$

From (5.31), we have

$$\delta H = \psi(2H^2 - K) + \frac{1}{2}g^{ij}\nabla_i\psi_j, \qquad (5.40)$$

which with (5.32) becomes

$$\delta H = \psi(2H^2 - K) + \frac{1}{2}g^{ij}(\psi_{ij} - \Gamma_{ij}^k\psi_k). \qquad (5.41)$$

Inserting (5.37) and (5.41) into (5.39), integrating ψ_{ij} and ψ_k by parts, and using these results, Eqs. (5.38) and (5.37), we obtain

$$\delta F = \oint \psi \left(\Delta p - 2\lambda H + k_c (2H + \mathbb{h})(2H^2 - 2K - \mathbb{h}H) \right.$$
$$\left. + (k_c / \sqrt{g})(\partial_i \partial_j + \partial_k \Gamma_{ij}^k) g^{ij} (2H + \mathbb{h}) \sqrt{g} \right) \mathrm{d}A. \qquad (5.42)$$

Additionally, by using Eqs. (5.17) and (5.18), one may prove the identity

$$\partial_i \left((\partial_i g^{ij} \sqrt{g}) f \right) = -\partial_k (\Gamma_{ij}^k g^{ij} \sqrt{g} f), \qquad (5.43)$$

in which $f(u, v)$ is an arbitrary function. Transforming the last term in Eq. (5.42) appropriately, we have

$$\delta F = \oint \psi \left(\Delta p - 2\lambda H + k_c (2H + \mathbb{h})(2H^2 - 2K - \mathbb{h}H) + 2k_c \Delta_S H \right) \mathrm{d}A, \qquad (5.44)$$

where Δ_S (in front of H) is the Laplace-Beltrami operator of the surface S, i.e.,

$$\Delta_S = (1/\sqrt{g}) \partial_i (g^{ij} \sqrt{g} \partial_j). \qquad (5.45)$$

Since $\mathbf{x}(u, v)$ describes an equilibrium shape, and if $\delta F = 0$ is satisfied for any infinitesimal function $\psi(u, v)$, Eq. (5.44) reduces to the equilibrium condition Ou-Yang and Helfrich (1989)

$$2k_c \Delta_S H - 2\lambda H + k_c (2H^2 - 2K - \mathbb{h}H)(2H + \mathbb{h}) + \Delta p = 0. \qquad (5.46)$$

This equation accounts for the balance of the normal forces per unit area. By contrast with the Laplace-Young equation, it contains, apart from each other, the pressure difference Δp and the tensile stress λ, a complicated expression due to the elasticity stresses.

5.5 Symmetries of the Shape Equation

In many cases in mathematical physics, the symmetry group of the problem under consideration is not obvious, and this suggests the task for its determination. Let us mention that if we know the symmetry group, this allows the set of solutions to be equiped with a linear structure which can be used for different purposes. In the next section, we will solve the above problem for the case of the shape equation derived in the previous one.

This will be done in Monge and conformal coordinates.

5.5.1 Cartesian Coordinates

In this section, we will consider the symmetries of the shape equation (5.46). Using the standard approach (see Ovsiannikov 1982; Ibragimov 1985; Bluman and Kumei

1989 and Olver 1993) for group analysis of differential equations, we can prove that in Monge parameterization (see Example 1.10), the symmetries of (5.46) coincide with the full group of motions in $\mathbb{R}^3$. The generators and their characteristics are represented in Table 5.1.

It was proven (see Vassilev et al. 2006) that all symmetries of Eq. (5.46) are variational as well, i.e., they are symmetries of the functional

$$\mathcal{F} = \mathcal{F}_c + \lambda \int_S \mathrm{d}A + p \int \mathrm{d}V, \qquad (5.47)$$

in which λ and p are real constants, that denote the surface tension and the osmotic pressure difference on both sides (inside and outside) of the membrane, and $\mathrm{d}V$ is the volume element. Respectively, in accordance with Noether's theorem, there are six linearly independent conservation laws

$$D_\alpha P_j^\alpha = Q_j E(L), \qquad \alpha = 1, 2, \qquad j = 1, \ldots, 6,$$

where

$$P_j^\alpha = N_j^\alpha L$$

$$N_j^\alpha = \xi_j^\alpha + Q_j \frac{\partial}{\partial w_\alpha} - \frac{1}{2} Q_j D_\mu \frac{\partial}{\partial w_{\alpha\mu}} - \frac{1}{2} Q_j D_\mu \frac{\partial}{\partial w_{\mu\alpha}}$$
$$+ \frac{1}{2} \left(D_\mu Q_j \right) \frac{\partial}{\partial w_{\alpha\mu}} + \frac{1}{2} \left(D_\mu Q_j \right) \frac{\partial}{\partial w_{\mu\alpha}}$$

$$D_\alpha = \frac{\partial}{\partial x^\alpha} + w_\alpha \frac{\partial}{\partial w} + w_{\alpha\mu} \frac{\partial}{\partial w_\mu} + \cdots$$

$$E = \frac{\partial}{\partial w} - D_\mu \frac{\partial}{\partial w_\mu} + D_\mu D_\nu \frac{\partial}{\partial w_{\mu\nu}} - \cdots.$$

	Generators	Characteristics
Table 5.1 Generators and characteristics of the group of motions in $\mathbb{R}^3$	*Translations*	
	$\mathbf{v}_1 = \frac{\partial}{\partial x^1}$	$Q_1 = -w_1$
	$\mathbf{v}_2 = \frac{\partial}{\partial x^2}$	$Q_2 = -w_2$
	$\mathbf{v}_3 = \frac{\partial}{\partial w}$	$Q_3 = 1$
	Rotations	
	$\mathbf{v}_4 = -x^2 \frac{\partial}{\partial x^1} + x^1 \frac{\partial}{\partial x^2}$	$Q_4 = x^2 w_1 - x^1 w_2$
	$\mathbf{v}_5 = -w \frac{\partial}{\partial x^1} + x^1 \frac{\partial}{\partial w}$	$Q_5 = x^1 + w w_1$
	$\mathbf{v}_6 = -w \frac{\partial}{\partial x^2} + x^2 \frac{\partial}{\partial w}$	$Q_6 = x^2 + w w_2$

5.5.2   Group-Invariant Solutions

It is easy to check that the vector fields

$$\langle \mathbf{v}_1 \rangle, \qquad \langle a\mathbf{v}_3 + \mathbf{v}_4 \rangle,$$

where a is a real constant, form an optimal system of one-dimensional subalgebras of the algebra of the symmetry group of the shape equation. Therefore, the different group-invariant solutions to Eq. (5.46) correspond to the vector fields $\mathbf{v}_1$ and $a\mathbf{v}_3 + \mathbf{v}_4$, and we can obtain every group-invariant solution through them. The two types of reduced group-invariant equation which determine the above group-invariant solutions are

(1) $G(\mathbf{v}_1)$ – invariant solution of the type

$$w = W(x^1)$$

of the reduced equation

$$\frac{k_c}{(v^2+1)^{5/2}} v_{11} - \frac{5k_c v}{2(v^2+1)^{7/2}} v_1^2 - \frac{1}{2}\frac{(k_c h^2 + 2\lambda) v}{(v^2+1)^{1/2}} + px^1 = C_1, \quad (5.48)$$

in which C_1 is a real number, $v = dW/dx^1$, $v_1 = dv/dx^1$ and $v_{11} = dv_1/dx^1$.
(2) $G(a\mathbf{v}_3 + \mathbf{v}_4)$–invariant solution of the type

$$w = \widehat{w}(r) + a\arctan(\frac{x^2}{x^1}), \qquad r = \sqrt{(x^1)^2 + (x^2)^2}$$

of the reduced equation written in the form

$$E_r(L_2) + pr = C_2,$$

where

$$E_r = \frac{\partial}{\partial v} - \left(\frac{\partial}{\partial r} + v_1\frac{\partial}{\partial v} + v_{11}\frac{\partial}{\partial v_1}\right)\frac{\partial}{\partial v_1}$$

$$L_2 = \frac{k_c F^2}{G^5} + \frac{2k_c h F}{G^2} + (k_c h^2 + 2\lambda) G$$

$$G^2 = r^2(v^2+1) + a^2, \qquad F = r(r^2 + a^2)v_1 + (G^2 + a^2)v,$$

and additionally, one should keep in mind that C_2 is a real number, $v = d\widehat{w}/dr$, and $v_1 = dv/dr$.

5.5.3 Conformal Coordinates

The shape equation in the Monge representation is a fourth order nonlinear partial differential equation. If we turn to a conformal metric on the plane and change the variables appropriately, the derivatives in the equation decrease their order. Simultaneously, we have to add three more equations–the so-called compatibility conditions of G-C-M (see Eqs. (1.77)). In general terms, this is the price that we have to pay for lowering the order of the equation. In this setting, we face the problem of finding the solutions to a system of four second order partial differential equations about four functions of two independent variables (for additional information, see de Matteis (2003) and Pulov et al. (2012).

It turns out that it is convenient to choose the conformal coordinates in which the metric and matrix h of the second fundamental form $h_{ij}dx^i dx^j$ (the repeated indices means that we have summation) take the form

$$ds^2 = g_{ij}dx^i dx^j = 4q^2\varphi^2(dx^2 + dz^2), \quad i, j = 1, 2, \quad x^1 = x, \ x^2 = z$$

$$h = \begin{pmatrix} \theta & \omega \\ \omega & 8q^2\varphi(1 + \hbar\varphi) - \theta \end{pmatrix}.$$

Here, q, φ, θ and ω are smooth functions of conformal coordinates x and z, and $\hbar$ is the spontaneous curvature which is connected with the mean curvature H by the formula

$$H = \frac{1}{\varphi} + \hbar. \tag{5.49}$$

Applying Brioschi's formula for the Gausian curvature

$$K = -\Delta \log(2q\varphi), \tag{5.50}$$

in which the explicit form (cf. formula (5.45)) of the Laplace-Beltrami operator Δ is

$$\Delta = \frac{1}{4q^2\varphi^2} \left(\frac{\partial^2}{\partial x^2} + \frac{\partial^2}{\partial z^2} \right), \tag{5.51}$$

we have

$$K = \frac{1}{4q^4\varphi^2}(q_x^2 + q_z^2) + \frac{1}{4q^2\varphi^4}(\varphi_x^2 + \varphi_z^2) - \frac{\Delta q}{q} - \frac{\Delta \varphi}{\varphi}.$$

The change in the variables in the shape equation results in a system of four partial differential equations

$$q^2 \Delta\varphi + 2q\varphi\Delta q - \frac{1}{2q^2\varphi}(q_x^2 + q_z^2) + \frac{q^2}{4\varphi^2}(8\varphi + \alpha_2\varphi^2 + \alpha_3\varphi^3 + \alpha_4\varphi^4) = 0$$

$$\theta_z - \omega_x - (8 + \frac{\alpha_2}{3}\varphi)q(\varphi q_z + q\varphi_z) = 0$$

(5.52)

$$\omega_z + \theta_x - \frac{\alpha_2}{3}q\varphi(\varphi q_x + q\varphi_x) - 8q\varphi q_x = 0$$

$$4q^2\varphi\Delta\varphi + 4q\varphi^2\Delta q - \frac{q_x^2 + q_z^2}{q^2} - \frac{\varphi_x^2 + \varphi_z^2}{\varphi^2} + (2 + \frac{\alpha_2}{12}\varphi)\frac{\theta}{\varphi} - \frac{\omega^2 + \theta^2}{4q^2\varphi^2} = 0,$$

in which

$$\alpha_2 = 24\text{h}, \qquad \alpha_3 = 8(2\text{h}^2 - \frac{\lambda}{k}), \qquad \alpha_4 = \frac{4p}{k} - \frac{8\lambda\text{h}}{k}.$$

5.5.4 Lie Equations

The principle aim of the group analysis of any system of differential equations is to find the maximal group of continuous transformations (the Lie group of point transformation) of dependent and independent variables for which this system is invariant. This is the so-called symmetry group (admissible group) of the system of differential equations. Since every Lie group is generated by its one-parameter subgroups, the problem is to find all groups of one-parameter transformations which keep the system unchanged. The one-parameter groups are defined by the system of the so-called Lie equations

$$\frac{d\Phi^i}{d\varepsilon} = \xi^i(\Phi, \Psi), \qquad \tilde{x}^i = \Phi^i(x, u, \varepsilon), \qquad \Phi^i|_{\varepsilon=0} = x^i, \qquad i = 1, 2$$

(5.53)

$$\frac{d\Psi^\alpha}{d\varepsilon} = \eta^\alpha(\Phi, \Psi), \qquad \tilde{u}^\alpha = \Psi^\alpha(x, u, \varepsilon), \qquad \Psi^\alpha|_{\varepsilon=0} = u^\alpha, \qquad \alpha = 1, 2, 3, 4$$

generating the flow of the vector field

$$X = \xi^i(x, u)\frac{\partial}{\partial x^i} + \eta^\alpha(x, u)\frac{\partial}{\partial u^\alpha}$$

(5.54)

in the space of independent and dependent variables (x, u). Here, ε is the parameter of the group ($\varepsilon \in I \subset \mathbb{R}, 0 \in I$), and x, u, Φ and Ψ are, respectively, vectors with coordinates x^i, u^α, Φ^i and Ψ^α.

The vector X is the so-called infinitesimal operator or generator of the symmetry group. Its second prolongation

$$\text{pr}^{(2)} X \equiv \tilde{X} = X + \eta_\alpha^x \frac{\partial}{\partial u_x^\alpha} + \eta_\alpha^z \frac{\partial}{\partial u_z^\alpha} + \eta_\alpha^{xx} \frac{\partial}{\partial u_{xx}^\alpha} + \eta_\alpha^{xz} \frac{\partial}{\partial u_{xz}^\alpha} + \eta_\alpha^{zz} \frac{\partial}{\partial u_{zz}^\alpha}$$

is the necessary ingredient for application of the Lie group techniques for analysis of the systems of differential equations. The coefficients η_α^x, η_α^{xx}, ... are expressed by the first and second order derivatives of $\xi^i(x, u)$, $u^\alpha(x, u)$ and $\eta^\alpha(x, u)$.

5.5.5 Determining System of Equation

At the first stage of group analysis, we create and solve the so-called determining system of equations

$$\tilde{X}[F_\nu] = 0, \qquad F_\nu = 0, \qquad \nu = 1, \ldots, 4, \tag{5.55}$$

where F_ν, $\nu = 1, \ldots, 4$, denote the expressions on the left-hand side of the system of Eq. (5.52). The huge number of equations in the determining system is the cause of serious technical difficulties. These difficulties can be at least partially surmounted by using some of the contemporary computer algebra systems. For the system (5.52), we have used the program *LieSymm-PDE* (see Pulov et al. 2007). In this way, we have obtained a determining system consisting of 206 partial differential equations.

After 17 contiguous executions of the program in interactive mode, the determining system was solved. The additional program modules were created, each of them solving a special kind of algebraic or differential equations. After each restart of the program, one introduces additional data—and these are either some algebraic or differential consequences of the equations that have yet to be solved or some solutions of them. At the end of each stage of the program, the functions ξ^i and η^α have a new form which approaches the desired solution—the generator of the symmetry group (5.52). The number of equations in the determining system decrease—after the third execution of the program, this number is 46, then it stays at 30, and quickly reduces to zero after that (see Pulov et al. 2012).

At the end, for infinitesimal operator $\bar{X}$ of the system of point symmetry group (5.52), we have the expression

$$\bar{X} = X_1 - q\xi_x^1 \frac{\partial}{\partial q} - 2(\theta\xi_x^1 + \omega\xi_x^2)\frac{\partial}{\partial\theta} - 2\left(\omega\xi_x^1 - \left(\theta - 4q^2\varphi - \frac{\alpha_2}{6}q^2\varphi^2\right)\xi_x^2\right)\frac{\partial}{\partial\omega},$$

where $X_1 = \xi^1 \frac{\partial}{\partial x} + \xi^2 \frac{\partial}{\partial z}$, and $\xi^1 = \xi^1(x, z)$ $\xi^2 = \xi^2(x, z)$ are arbitrary functions, satisfying the Cauchy–Riemann condition

$$\frac{\partial\xi^1}{\partial z} = -\frac{\partial\xi^2}{\partial x}, \qquad \frac{\partial\xi^1}{\partial x} = \frac{\partial\xi^2}{\partial z}. \tag{5.56}$$

Remark

The vector field $\bar{X}$ is the generator of the symmetry group (5.52) for $\alpha_k \neq 0$, $k = 2, 3, 4$. In the case in which $\alpha_k \equiv 0$ (equivalent to the Willmore's equation), the generator of the admissible group is

$$\hat{X} = \bar{X} + C \left(\varphi \frac{\partial}{\partial \varphi} + \theta \frac{\partial}{\partial \theta} + \omega \frac{\partial}{\partial \omega} \right), \qquad C = \text{const.}$$

References

G. Bluman, S. Kumei, *Symmetries and Differential Equations* (Springer, New York, 1989)

B. Canham, The minimum energy of bending as a possible explanation of biconcave shape of the human red blood cell. J. Theoret. Biol. **26**, 61–81 (1970)

H. Davson, J. Danielli, *The Permeability of Natural Membranes*, 2nd edn. (Cambridge University Press, Cambridge, 1952)

G. de Matteis, Group analysis of the membrane shape equation, in *Nonlinear Physics: Theory and Experiment II*, ed. by M. Ablowitz, M. Boiti, F. Pempinelli, B. Prinari (World Scientific, Singapore, 2003), pp. 221–226

H. Deuling, W. Helfrich, Red blood cell shapes as explained on the basis of curvature elasticity. Biophys. J. **16**, 861–868 (1976)

Y.-C. Fung, P. Tong, Theory of the sphering of red blood cell. Biophys. J. **8**, 175–198 (1968)

W. Helfrich, Elastic properties of lipid bilayers: Theory and possible experiments. Z. Naturforsch c **28**, 693–703 (1973)

W. Helfrich, J. Deuling, Some theoretical shapes of red blood cells. J. Phys. (Colloq.) **36**, 327–329 (1975)

N. Ibragimov, *Transformation Groups Applied to Mathematical Physics* (Riedel, Boston, 1985)

T. Jenkins, Static equilibrium configurations of a model red blood cell. J. Math. Biol. **4**, 149–169 (1977a)

T. Jenkins, The equations of mechanical equilibrium of a model membrane. SIAM J. Appl. Math. **32**, 755–764 (1977b)

J. Murphy, Erythrocyte metabolism VI., Cell shape and the location of cholesterol in the erythrocyte membrane. J. Lab. Clin. Med. **65**, 756–774 (1965)

P. Olver, *Applications of Lie Groups to Differential Equations*, 2nd edn. (Springer, New York, 1993)

Z.-C. OuYang, W. Helfrich, Bending energy of vesicle membranes: General expressions for the first, second, and third variation of the shape energy and applications to spheres and cylinders. Phys. Rev. A **39**, 5280–5288 (1989)

L. Ovsiannikov, *Group Analysis of Differential Equations* (Academic, New York, 1982)

V. Pulov, M. Hadzhilazova, V. Lyutskanov, I. Mladenov, Helfrich's shape model of biological membranes: Group analysis, determining system and symmetries, in *Proceedings of the Third International Scientific Congress* **7** (TU Varna, 2012), pp. 271–275

V. Pulov, E. Chacarov, I. Uzunov, A computer algebra application to determination of Lie symmetries of partial differential equations. Serdica J. Comput. **1**, 241–251 (2007)

R. Rand, Mechanical properties of the red cell membrane, II. Viscoelastic breakdown of the membrane. Biophys. J. **4**, 303–316 (1964)

R. Rand, A. Burton, Mechanical properties of the red cell membrane, I. Membrane stiffness and intracellular pressure. Biophys. J. **4**, 115–135 (1964)

M. Schneider, J. Jenkins, W. Webb, Thermal fluctuations of large cylindrical phospholipid vesicles. Biophys. J. **45**, 891–899 (1984)

V. Vassilev, P. Djonjorov, I. Mladenov, Symmetry groups, conservation laws and group-invariant solutions of the membrane shape equation. Geom. Integrab. Quant. **7**, 265–279 (2006)

Chapter 6
Exact Solutions and Applications

Abstract This chapter considers the construction of the few known and some new explicit solutions to the shape equation. For example, a review of the available experimental facts concerning nerve fibers suggests that their form can be adequately modelled by Delaunay unduloids. We present parameterizations of these surfaces in terms of the geometrical parameters involved in the model. This allows for direct expression of all geometric characteristics to the membranes, like the length of any separate extension, their volume and surface area via explicit formulas. This in turn allows for the direct examination of sensitivity of these characteristics on the amplitude oscillations of the parameters and animation of the morphological changes. Note that this is either impossible or very difficult to achieve by means of numerical analysis. The same applies to the modelling of the results in Cole's experiments, which can be described in terms of Delaunay's nodoids. Besides analytical formulas, we consider the geometry of the egg of the sea urchin under bilateral deformation between two plates of the model and determine the surface tension of the membrane of the egg as a direct consequence of analytical results. Another instance when Delaunay's nodoids appear quite naturally is in the modelling of the membrane fusion. According to the commonly-accepted assumption, this process involves an hour-glass-shaped local contact between two monolayers of opposing membranes, an intermediate structure which is called a stalk. The shape of the stalk is considered to be an axisymmetrical figure of revolution in the three-dimensional space with a planar geometry in the initial configuration. The total energy of the stalk is evaluated from the assumption that the stalk is a nodoid with constant curvature. The conclusions of the group analysis in the previous chapter are used to distinguish two classes of analytical solution of the shape equation, the translationally-invariant and axially-symmetric, to be precise. In the first case, the Ou-Yang equation is reduced to the equation for a so-called generalized elastica, whose solutions are fully described and illustrated with graphics. The second class of axially-symmetric membranes is the family of Delaunay-like surfaces, but with non-constant mean curvatures. These surfaces constitute the first examples of a surfaces with periodic mean curvatures.

© Springer International Publishing AG 2017
I.M. Mladenov and M. Hadzhilazova, *The Many Faces of Elastica*,
Forum for Interdisciplinary Mathematics 3, DOI 10.1007/978-3-319-61244-7_6

6.1   Unduloids and Nerve Fibers

6.1.1   Introduction

Myelinated and unmyelinated nerve fibers, which are subjected to stretch, exhibit beaded shapes, resulting in a succession of expansions and constrictions observed along the axis of longitudinal sections of the fibres, as can be seen in Figs. 6.1 and 6.2.

Various explanations of the mechanisms responsible for the beading of nerve fibres and cultured neurites have been proposed in the literature. Exploring these works reveals that beading may result from the action of various stimuli: mechanical, as described in Ochs and Jersild (1990), electrical, per Manthy et al. (1995), chemical, via a metabolic perturbation used by Tanelian and Markin (1997), or a laser-induced mechanical force, as used by Bar-Ziv and Moses (1994). It can also simply be viewed as a normal physiological process caused by activity of the nervous system (Allen et al. (1997) and Zhu et al. (1997)).

The constrictions cause compaction of the cytoskeleton and its enclosed organelles, including microtubules and neurofilaments, which become more closely apposed. One of the mechanisms conceived for beading is the redistribution of the membrane bilayer, which is free to rearrange due to the cytoskeleton disaggregation, as pointed out by Tanelian and Markin (1997). In this model, the final shape is approximated by a succession of spheres and cylinders. The model suffers, however, from the drawback

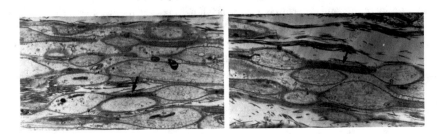

Fig. 6.1 Beaded axons in unmyelinated fibers. A succession of expansions and constrictions can be seen within the axons. Note the compacted cytoskeleton within the constrictions (*arrow*), causing them to appear as a dark-stained band. Magnification ×11,000. (Markin et al. 1999)

Fig. 6.2 A single myelinated fiber teased from a rat sciatic nerve stretched with a weight of 4.5 g, to bead its fibers and cold-fixed while remaining under stretch (Markin et al. 1999)

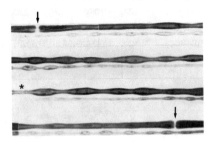

that the final shape of the beading is postulated, instead of being calculated. Beaded shapes can indeed differ according to the compaction of the cytoskeleton within the constrictions in different types of nerve fibres (Ochs et al. 1997). A more general model which was able to predict the universal shapes of beaded fibres, based on the deformation of the axonal membrane under the action of hydrostatic pressure and tension, was proposed in Markin et al. (1999). The main determinant for beading in this more recent model is the mechanical response of the fibre membrane, relying on observations of the constricted regions of the unmyelinated nerve fibre (Ochs et al. 1996), in which the cytoskeleton is not depolymerised as imagined by Tanelian and Markin (1997), but instead becomes compressed together with its organelles into a packed constricted region.

6.1.2 Model

The stretching force applied to the membrane edge generates tension in the membrane, which in turn creates hydrostatic pressure in the cytoplasm. Since a fluid surface with surface tension is unstable, it will change its shape, according to the Laplace–Young law. According to the Laplace–Young equation (4.1), we can write

$$\Delta p = \sigma(\kappa_\pi + \kappa_\mu), \tag{6.1}$$

where Δp is the transmembrane pressure differential, σ is the surface tension, and κ_π and κ_μ are, respectively, the so-called parallel, and meridional principal curvatures (see (1.80)) of the axially symmetric surface presenting the beaded shape. The cylindrical part of the transformed nerve has a larger mean curvature than the beaded region. According to Eq. (6.1), the membrane over the cylindrical region produces a higher pressure, counteracted by an opposing pressure created by the compressed cytoskeleton core. From a geometrical viewpoint, two beads are generally separated by a compacted neck, but the limit situation of directly connected beds is also possible.

6.1.3 Parameterization

In Markin et al. (1999), the nerve fibre is modelled as a cylindrical fluid membrane having the ability to change its shape easily at constant membrane area and fibre volume. Furthermore, since the ratio of Δp to σ is also a constant, the Laplace–Young equation can be written as

$$\kappa_\pi + \kappa_\mu = \text{const}, \tag{6.2}$$

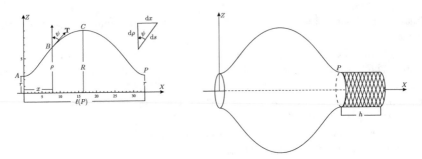

Fig. 6.3 Geometry of the profile curve of Delaunay's unduloid (*left*) and that of a full segment of the periodic beaded shape (*right*)

revealing the fact that we are dealing here with the class of the so-called Delaunay surfaces described in Sect. 4.3 (for more details, see Eells (1987) and Oprea (2007)). Kenmotsu (1980) had shown that rotational surfaces of a given mean curvature in $\mathbb{R}^3$ are defined essentially by their Gauss map. Later on, Eells (1987) pointed out that the Gauss map for the Delaunay surfaces is given by the general formula

$$\sin \psi(\rho) = m\rho + \frac{n}{\rho}, \qquad \rho \neq 0, \qquad m, n \in \mathbb{R}, \tag{6.3}$$

where ρ is the distance from the symmetry axis, m and n are real parameters, and $\psi(\rho)$ is the angle between the tangent **T** to the profile curve at the current point B and the Z axis (cf. Fig. 6.3). Introducing $\rho_{\max} = R$ and $\rho_{\min} = r$, we easily find that in our case, m and n are given by the formulas

$$m = \frac{1}{R+r}, \qquad n = \frac{Rr}{R+r}, \tag{6.4}$$

and that the meridional curve is determined by the equation

$$\frac{\mathrm{d}x}{\mathrm{d}\rho} = \tan \psi(\rho) = \frac{\rho^2 + Rr}{\sqrt{(R^2 - \rho^2)(\rho^2 - r^2)}}. \tag{6.5}$$

The integration of (6.5) amounts to evaluating the integral

$$\int \frac{(\rho^2 + Rr)\mathrm{d}\rho}{\sqrt{(R^2 - \rho^2)(\rho^2 - r^2)}}, \tag{6.6}$$

which can be performed via the Jacobian elliptic function $\mathrm{dn}(u, k)$, i.e.,

$$\rho = \frac{r}{\mathrm{dn}(u, k)}, \qquad k = \frac{\sqrt{R^2 - r^2}}{R}, \tag{6.7}$$

where u is its argument and k is the so-called elliptic modulus. The above substitution immediately produces

$$\frac{d\rho}{\sqrt{(R^2 - \rho^2)(\rho^2 - r^2)}} = \frac{du}{R}, \tag{6.8}$$

and therefore

$$x(u) = \frac{1}{R} \int \rho^2(u)du + ru. \tag{6.9}$$

The first integral can be evaluated by taking into account the formula (cf. formula 315.02 in Byrd and Friedman (1971))

$$\int \frac{du}{dn^2(u, k)} = \frac{1}{\tilde{k}^2} \left(E(am(u, k), k) - k^2 \frac{sn(u, k) cn(u, k)}{dn(u, k)} \right), \tag{6.10}$$

in which $\tilde{k}$ is the so-called complementary elliptic modulus

$$\tilde{k}^2 = 1 - k^2 = \frac{r^2}{R^2}, \tag{6.11}$$

while the second one is trivial. Taken together, they give us

$$x(u) = RE(am(u, k), k) + rF(am(u, k), u) - Rk^2 \frac{sn(u, k)cn(u, k)}{dn(u, k)}. \tag{6.12}$$

6.1.4 Parameters of the Nerve Fibers

The length $\ell(P)$ of a single bead (Fig. 6.3) can be found by noticing that the real period of $dn(u, k)$ is $2K(k)$, along with the equality $am(K(k), k) = \pi/2$ and (6.12), which, combined, lead to the result

$$\ell(P) = 2(RE(k) + rK(k)). \tag{6.13}$$

The area of the striped part of the undulated surface (see Fig. 6.4) is given by the application of the slice formula

$$A(B) = 2\pi \int \rho ds, \tag{6.14}$$

where

$$ds = \sqrt{1 + \left(\frac{dx}{d\rho}\right)^2} \, d\rho = \frac{(R + r)\rho d\rho}{\sqrt{R^2 - \rho^2)(\rho^2 - r^2)}}, \tag{6.15}$$

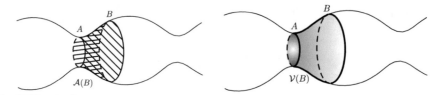

Fig. 6.4 The area of the striped part of the unduloid surface is given by formula (6.17). The *dotted* domain presents the volume enclosed by the unduloid surface and the two disks through the points A and B

and therefore

$$A(B) = 2\pi(R + r) \int \frac{\rho^2 d\rho}{\sqrt{(R^2 - \rho^2)(\rho^2 - r^2)}}. \tag{6.16}$$

The integration is immediate due to formula (6.10) and gives,

$$A(B) = 2\pi(R + r)R \left(E(\text{am}(u, k), k) - k^2 \frac{\text{sn}(u, k)\text{cn}(u, k)}{\text{dn}(u, k)} \right). \tag{6.17}$$

Obviously, the surface area of the whole bead is

$$A(P) = 2A(C) = 4\pi R(R + r)E(k). \tag{6.18}$$

The volume enclosed by the unduloid surface and the two disks through the points A and B can be found in a similar way. Together, the slice formula (6.5) and (6.7) lead to the following result:

$$\begin{aligned}
V(B) = \quad & \pi \int_0^u \rho^2 dx = \pi \int_0^u \frac{\rho^2(\rho^2 + Rr)d\rho}{\sqrt{R^2 - \rho^2)(\rho^2 - r^2)}} \\
= \quad & \pi \left(\frac{r^4}{R} \int_0^u \frac{du}{\text{dn}^4(u, k)} + r^3 \int_0^u \frac{du}{\text{dn}^2(u, k)} \right).
\end{aligned} \tag{6.19}$$

The formula for the first of the integrals above is also available, namely (see Byrd and Friedman (1971), formula 315.04),

$$\begin{aligned}
\int \frac{du}{\text{dn}^4(u, k)} = \frac{1}{3\tilde{k}^4} \Bigg[& 2(2 - k^2) \left(E(\text{am}(u, k), k) - k^2 \frac{\text{sn}(u, k)\text{cn}(u, k)}{\text{dn}(u, k)} \right) \\
& -\tilde{k}^2 F(\text{am}(u, k), k) - k^2 \tilde{k}^2 \frac{\text{sn}(u, k)\text{cn}(u, k)}{\text{dn}^3(u, k)} \Bigg],
\end{aligned}$$

and having (6.10) in mind, we end up with the expression

$$V(B) = \frac{\pi R}{3}[(2R^2 + 3Rr + 2r^2)E(\text{am}(u, k), k) - r^2 F(\text{am}(u, k), k)$$

$$-(2R^2 + 3Rr + 3r^2)k^2 \frac{\text{sn}(u, k)\text{cn}(u, k)}{\text{dn}(u, k)}]. \tag{6.20}$$

The formula for the volume of the entire bead follows immediately

$$V(P) = \frac{2\pi R}{3}[(2R^2 + 3Rr + 2r^2)E(k) - r^2 K(k)]. \tag{6.21}$$

6.1.5 Sensitivity of the Equilibrium Shapes on the Parameters

The surface and volume enclosed by the membrane are assumed to remain unchanged during the transformation of the nerve fibre caused by the action of hydrostatic pressure and axial tension. As the initial configuration is a cylinder of radius $\mathring{r}$ and length ℓ_0, we have

$$A = 2\pi \mathring{r}\ell_0, \quad V = \pi \mathring{r}^2 \ell_0, \quad \frac{V}{A} = \frac{\mathring{r}}{2} = \frac{V(P)}{A(P)},$$

and therefore

$$\mathring{r} = 2\frac{V(P)}{A(P)}. \tag{6.22}$$

By definition, the average radius $\bar{r}$ of the bead is

$$\bar{r} = \frac{R + r}{2}, \tag{6.23}$$

which allows us to write down the following relations:

$$\frac{\ell(P)}{\bar{r}} = \frac{4(RE(k) + rK(k))}{R + r}, \quad \frac{\ell(P)}{\mathring{r}} = \frac{(RE(k) + rK(k))A(P)}{V(P)}$$

$$\frac{\bar{r}}{\mathring{r}} = \frac{R + r}{4}\frac{A(P)}{V(P)}. \tag{6.24}$$

As a result of stretching, the length of the fibre, in principal, should increase, and it is an interesting problem to analyze how this deformation actually depends on the oscillating amplitude a defined by the formula

$$a = \frac{R - r}{R + r}. \tag{6.25}$$

Corresponding plots of the dependencies of (6.24) on a are presented in Fig. 6.5.

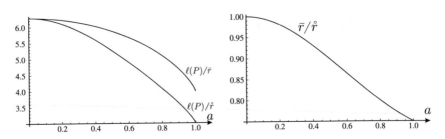

Fig. 6.5 The plots of $\ell(P)/\bar{r}$ and $\ell(P)/\mathring{r}$ (*left*), and that of the average radius normalized by $\mathring{r}$ (*right*)

It is also immediately possible to prove that we have

$$r = \frac{1-a}{1+a}R = (1-a)\bar{r}, \qquad k = \frac{2\sqrt{a}}{1+a},$$

which makes clear that the shape of the bead depends exclusively on the amplitude, while R specifies its bulk size

$$\ell = A\frac{L_{\text{tot}}}{A_{\text{tot}}} = 2\pi\mathring{r}\ell_0\frac{L_{\text{tot}}}{A_{\text{tot}}}.$$

The latter means that we can find, respectively, the ratio

$$\frac{\ell}{\ell_0} = 4\pi\frac{\mathcal{V}(P)}{\mathcal{A}(P)}\frac{L_{\text{tot}}}{A_{\text{tot}}} = 4\pi\frac{\mathcal{V}(P)}{\mathcal{A}(P)}\frac{\ell(P)+h}{\mathcal{A}(P)+2\pi rh},$$

and the elongation

$$\epsilon = \frac{\ell}{\ell_0} - 1.$$

Its variations as a function of a and h are depicted in Fig. 6.6.

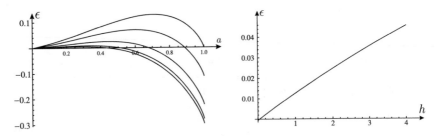

Fig. 6.6 The plots of elongation ϵ as a function of the amplitude oscillation a and different values of h (*left*), and that of the length h of the cylindrical neck (*right*) for the fixed value of $a = 0.2$

The evolution recorded in Fig. 6.6 shows that the elongation is limited to values less than 15% in tension, but can reach values up to 30% in compression (when $R \geq h$). A maximum in ϵ is observed for values of the parameter h higher than a certain critical value a.

The closed form expressions which have been obtained for the beaded shape of neurons under stretch represent a significant improvement compared to the Markin et al. (1999) approach based entirely on numerical solution of the determining equations. These results allow for a sensitivity analysis of the dependence of the equilibrium shapes on the model parameters and visualization of morphological transformations.

6.2 Mathematical Model of the Cole Experiment

The main aim in this section is to study in some detail the strongly nonlinear behavior of the vesicle deformations. The lipid membrane is treated as a thin elastic shell that possesses four modes of deformation—dilation, bending, shearing and torsion. From the geometrical viewpoint, the bending and torsion are related to the variations of the two principal curvatures of the interface.

The curvature dependence of the interfacial tension was first investigated by Young and Laplace. The variational problem is connected with the minimization of the functional

$$\sigma \int d\mathcal{A} + \Delta p \int dV \tag{6.26}$$

and leads to the Laplace–Young equation

$$\Delta p = 2\sigma H, \tag{6.27}$$

which can be easily established by relying on the results on the variations of the surface area and the volume, i.e.,

$$\delta \oint d\mathcal{A} = -2 \oint \psi H d\mathcal{A}, \qquad \delta \int dV = \int \psi d\mathcal{A}. \tag{6.28}$$

Having in mind (6.28), the variation of (6.26) results in

$$-2\sigma \oint \psi H d\mathcal{A} + \Delta p \int \psi d\mathcal{A}, \tag{6.29}$$

which is zero at a minimizer, i.e.,

$$\oint \psi (\Delta p - 2\sigma H) d\mathcal{A} = 0. \tag{6.30}$$

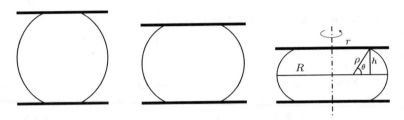

Fig. 6.7 Geometry of Cole's experiment

As the latter has to be satisfied for any smooth function ψ, this equality is equivalent to Eq. (6.27).

In the above proof, we have not considered any specific shape of the membrane (for example, with some symmetry) and consequently, Eq. (6.27) is valid for an arbitrary membrane.

The Laplace–Young equation (6.27), which describes the equilibrium between the surface pressure and surface tension, has been used by Cole (1932) in his experiments. In these experiments, the vesicles (sea-urchin eggs) have been compressed with known forces between two parallel plates and the surface tension was calculated by measuring the degrees of their flattening.

Later on, a similar experiment was performed by Yoneda (1964) who had, however, adopted another defining equation for σ.

Here, we will consider in some detail two different geometrical models which are relevant to the appropriate experiments. They have been chosen because they lead to relatively simple mathematical expressions (see Hadzhilazova and Mladenov (2004, 2005)) whose real merit is their analytical form.

6.2.1   Cole Model

It seems appropriate to recall here the essential points of Cole's method of calculation. Cole compressed a spherical egg of initial radius a between two parallel plates with a fixed force F. The surface tension was then evaluated by the measurement of flattening (the half-thickness h), the radius r of the contact area A and the equatorial radius R (see Fig. 6.7).

The values of the parameters h and R can be measured with high enough accuracy, but this is not the case with r, particularly when the contact angle is very close to 180°. The other two parameters which enter into the model are the inner radius ρ of the torus-like outer part of the membrane and the contact angle θ. Relying again on geometry, one can easily find that the above parameters are given by the expressions

$$\rho = \frac{h^2 + (R - r)^2}{2(R - r)}, \qquad \theta = \arcsin \frac{h}{\rho}. \tag{6.31}$$

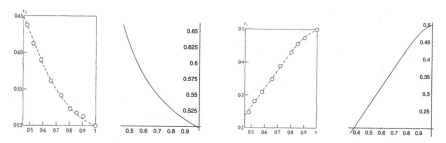

Fig. 6.8 The experimental results and the approximate curves for R are presented on the *left*, and those for ρ on the *right*

The profile curve of the torus-like part of the egg is parameterized explicitly by the formulas

$$x = R - \rho + \rho \cos u, \quad z = \rho \sin u, \quad u \in [-\arcsin \frac{h}{\rho}, \arcsin \frac{h}{\rho}],$$

and this is enough for find, respectively, the volume, and the surface area in the form (cf. Hadzhilazova and Mladenov (2004))

$$V = 2\pi[h(R^2 + 2\rho^2 - 2R\rho + (R - \rho)\sqrt{\rho^2 - h^2}) + \rho^2(R - \rho)\theta - \frac{h^3}{3}] \quad (6.32)$$

$$S = 4\pi\rho[(R - \rho)\theta + h] + 2\pi r^2. \quad (6.33)$$

Through photographically obtained values for R, ρ and h, the first two are plotted against the third and, after that, analytically fitted by explicit functions of $h \equiv z$. The remaining parameter r can be found by solving Eq. (6.31) with respect to this variable, and this gives us

$$r(z) = R - \rho - \sqrt{\rho^2 - z^2}.$$

Having $\rho(z)$, $R(z)$ and $r(z)$, one can put them back into (6.32) in order practically to check the constancy of the volume $V = (4/3)\pi a^3$. Besides, one can find the values of $1/R + 1/\rho$ for the points at the equator and, using

$$\Delta p = F/A = \sigma(1/R + 1/\rho), \quad (6.34)$$

determine the surface tension σ, which was the main purpose of the Cole (1932) experiment (Fig. 6.8).

6.2.2   *Yoneda Method*

In order to bypass the ambiguity in the measurement of the radius r of the contact disk area, Yoneda (1964) proposed another method for calculation of the surface tension. It is based on the following arguments.

Let us assume that the egg is compressed to the thickness z under the external force F. If under a slightly increasing force, the egg is compressed further by the distance $-\mathrm{d}z$, the work required for this additional compression is $-F\,\mathrm{d}z$, assumed to be expended entirely for stretching of the cortex by neglecting any other effects ((bending, etc.). If $\mathrm{d}S$ is the stretching of the surface produced by the compression, this work is just the surface tension σ (supposing that it is uniform along the entire surface) multiplied by $\mathrm{d}S$, i.e.,

$$ -F\mathrm{d}z = \sigma\mathrm{d}S \quad \text{or} \quad F = -\sigma\frac{\mathrm{d}S}{\mathrm{d}z}. \tag{6.35}$$

If F is plotted against $-\dfrac{\mathrm{d}S}{\mathrm{d}z}$, the latter equation implies that σ is given as the slope of the line through the origin, and this is confirmed experimentally by Yoneda (1964).

6.2.3   *Nodoids and the Compression of Spherical Eggs*

Looking at (6.34), it is clear that the surface tension will be constant provided we deal with axially symmetric surfaces of constant mean curvature, which, according to the Delaunay classification, are called nodoids. The profile curve of such a surface is periodic along the symmetry axis and has one local minimum and one local maximum in each period. Its geometry is presented at the outermost right-hand side in Fig. 6.9.

We can describe the deformed part of the egg in analytical form using the formulas

$$ x(u) = R\sqrt{1 - k^2 \sin^2 u}, \qquad z(u) = R\left(E(u,k) - \varepsilon^2 F(u,k)\right), \tag{6.36}$$

where

Fig. 6.9 Nodoid's geometry

Fig. 6.10 The graphics of
the function on the
right-hand side of Eq. (6.41)
and its linear fractional
approximation

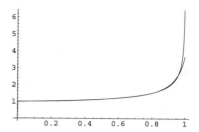

$$u \in [- \arcsin \frac{1}{\sqrt{1+\varepsilon^2}}, \arcsin \frac{1}{\sqrt{1+\varepsilon^2}}], \quad k^2 = 1 - \varepsilon^4, \quad \varepsilon = r/R. \quad (6.37)$$

Here, $F(u, k)$ and $E(u, k)$ denote the incomplete elliptic integrals of the first and second kinds, and k is the elliptic modulus. Using the explicit parameterization (6.36) of the compressed egg profile, we can prove that the infinitesimal arclength of the profile curve is determined by the formula

$$ds = \frac{R^2 + r^2}{R} du \quad (6.38)$$

and that the mean curvature of the deformed egg is

$$H = \frac{R}{r^2 - R^2}. \quad (6.39)$$

According to the second formula in (6.36), the height of the contact plane is

$$z(\varepsilon) = R \left(E(\arcsin \frac{1}{\sqrt{1+\varepsilon^2}}, k) - \varepsilon^2 F(\arcsin \frac{1}{\sqrt{1+\varepsilon^2}}, k) \right). \quad (6.40)$$

By again using the fact that during the compression the volume is conserved, one can find the relationship between the geometrical parameters that are needed - the initial radius a and that of the deformed egg R. The realization of this strategy gives us the relationship

$$\frac{R}{a} = \left[\left((1 - \frac{3\varepsilon^2}{2} + \varepsilon^4) (1 - \varepsilon^2 + E(k)) + \varepsilon^2(1 - \varepsilon^2 - \frac{\varepsilon^2 K(k)}{2}) \right)/2 \right]^{-\frac{1}{3}}, \quad (6.41)$$

and this means that knowing a, measuring R and using formula (6.41), one can find ε, and therefore, via (6.36) and (6.37), the profile curve of the compressed egg.

Taking into account that the right-hand side of (6.41) is a function only of the deformation parameter ε and that its values belong to the interval $0 \le \varepsilon \le 1$, we can find a linear fractional function which approximates it very well (see Fig. 6.10). We find, with accuracy exceeding the experimental accuracy, that we have the relations

$$R = a\frac{0.99098 - 0.77075\varepsilon}{1 - 0.93678\varepsilon}, \qquad \varepsilon = \frac{99098a - 100000R}{77075a - 93678R}, \qquad r = \varepsilon R. \quad (6.42)$$

The last equation means that it is not necessary to measure the radius of the contact area!

If one wants to use Yoneda's equation (6.35), it is necessary to find the area of the surface outside the contact area, which is defined by the formula

$$\mathcal{A}(\varepsilon) = 2\pi \int_0^{\arcsin\frac{1}{\sqrt{1+\varepsilon^2}}} x(u)ds = 2\pi R^2(1+\varepsilon^2)E\left(\arcsin\frac{1}{\sqrt{1+\varepsilon^2}}, k\right). \quad (6.43)$$

By adding the surface of the contact disk $\pi r^2 = \pi R^2\varepsilon^2$, one finds the full area (of the upper side) of the egg in the form

$$S(\varepsilon) = \pi R^2 \left(2(1+\varepsilon^2)E(\arcsin\frac{1}{\sqrt{1+\varepsilon^2}}, k) + \varepsilon^2\right). \quad (6.44)$$

We note that the expressions for the surface area (6.44) and the height (6.40) are analytical functions only of the deformation parameter ε and their differentials in Yoneda's formula (6.35) (Fig. 6.11).

For illustration, we present the differentiation result in graphic form, which should convince the reader that the surface tension is independent from the egg area, excluding the points near the origin. This is exactly Yoneda's conclusion, and his explanation is that this is because of experimental errors due to measuring small forces during slight compression. In some sense, this then proves that the model based on Delaunay's surfaces is correct.

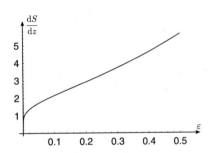

Fig. 6.11 The graphics of the derivative in Yoneda's equation (6.35)

6.3 Fusion of Membranes

6.3.1 Stalk Model

The fusion process involves the successive steps of membrane aggregation, a desta-
bilization nucleating at a point defect inducing a highly localized rearrangement of
the two bilayers, and a consequent mixing of the components of the two bilayers,
resulting in either hemifusion (Fig. 6.12, in the middle) or full fusion (Fig. 6.12, right).

The interaction of membranes involves local contacts between two phospholipid
bilayers in their aqueous environment, which is difficult due to the hydrophobic nature
of the interior part of the membranes (the bilayer has a trans and a cis monolayer,
Fig. 6.13). The required connection between the two membranes in the fusion process
involves an hourglass-shaped local contact between two monolayers of opposing
membranes, which is an intermediate structure called a stalk in the original model
developed in Kozlov and Markin (1983).

The mechanical basis of the model relies on calculation of the shape of the stalk,
taken as an axisymmetrical surface of revolution in 3D space (Fig. 6.13), with a
planar geometry in the initial configuration. The neutral surface is represented as a
dotted line. Here, x and z are the coordinates of the contour, the parameter a is the
shortest distance separating the neutral surface from the axis of revolution (the neck
of the stalk), c is the distance from the axis of revolution to the point where the stalk
branches become horizontal (the width of the stalk), $2h$ is the distance separating the
two neutral surfaces, and the angle between the neutral surface and the horizontal
axis OX is ψ.

Fig. 6.12 Schematic view of the process of fusion via stalk formation

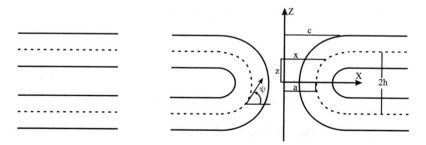

Fig. 6.13 Steps in membrane fusion. *Solid lines* represent hydrophilic surfaces, *dotted lines*
hydrophobic surfaces (*left*). Geometrical parameters of the stalk (*right*)

The contour of the stalk surface is given by the slope angle ψ. The bending energy of the stalk depends on the principal and the spontaneous curvatures. Assuming that the curvature of the stalk is constant, the total energy of the stalk is calculated versus the parameter a, and is found to be negative. Hence, it promotes hemifusion (the presence of spontaneous curvature in the monolayers favors hemifusion). The present section proposes an extension of the model suggested in Markin and Albanesi (2002), by considering explicitly a subclass of the Delaunay surfaces, the so-called nodoid surfaces. Following the argumentation developed in Hadzhilazova et al. (2010) in the case of beaded nerve fibers (the surfaces there are unduloids), we elaborate a model in which any geometrical characteristic of the stalks can be expressed in an explicit form.

6.3.2 Mathematical Description of the Stalk Model

The basic parameter in the model is the magnitude c_{stalk}, which is connected with the mean curvature of the stalk by the formula

$$c_p(x) + c_m(x) = c_{\text{stalk}}. \tag{6.45}$$

Using the fact that the surface is axially symmetric, the two principal curvatures may be expressed versus the angle $\psi(x)$ as

$$c_p(x) = \frac{\sin \psi(x)}{x}, \qquad c_m(x) = \cos \psi(x) \frac{d\psi}{dx}. \tag{6.46}$$

Therefore, the slope at any point along the contour of the stalk surface is determined from

$$\frac{dz}{dx} = \tan \psi(x) = \frac{x c_p}{\sqrt{1 - x^2 c_p^2}}. \tag{6.47}$$

Previous relations then lead, after straightforward calculations, to the equations

$$x c_p = \frac{1}{2} c_{\text{stalk}} x + \frac{a}{x}(1 - \frac{1}{2} c_{\text{stalk}} a) \tag{6.48}$$

$$\frac{dz}{dx} = \{[\frac{1}{2} c_{\text{stalk}} x + \frac{a}{x}(1 - \frac{1}{2} c_{\text{stalk}} a)]^{-2} - 1\}^{-1/2}. \tag{6.49}$$

Next, due to the above relations, we also have also the equations

$$c = \sqrt{a^2 - \frac{2a}{c_{\text{stalk}}}} \quad \text{and} \quad \frac{c}{a} = \sqrt{1 - \frac{2}{a c_{\text{stalk}}}}. \tag{6.50}$$

From all of the above, we see that the contour of the stalk surface is given as the integral

$$\frac{z}{a} = \int_1^{x/a} ([\frac{1}{2}ac_{\text{stalk}}t + \frac{1}{t}(1 - \frac{1}{2}ac_{\text{stalk}})]^{-2} - 1)^{-1/2}dt. \tag{6.51}$$

When $c_{\text{stalk}} = \mathrm{h} = \text{const}$, the stalk is called stress free. The surface specified by Eq. (6.45) is a constant mean curvature (CMC) surface. The CMC surfaces of revolution were classified a long time ago by the French geometer Delaunay (1841) and were described in analytical form in Mladenov (2002), Mladenov (2002a) and in Sect. 4.3. Differentiating $c_p(x)$ and taking into account the expression for $c_m(x)$, one gets the equation

$$\frac{dc_p(x)}{dx} = \frac{c_m(x) - c_p(x)}{x}. \tag{6.52}$$

This equation and the CMC condition

$$H = \frac{c_m(x) + c_p(x)}{2} = \mathrm{h} = \text{const} \tag{6.53}$$

yield

$$\frac{dc_p(x)}{dx} = \frac{2(\mathrm{h} - c_p(x))}{x}, \tag{6.54}$$

and therefore

$$c_p(x) = \mathrm{h} + \frac{\mathrm{b}}{x^2}, \tag{6.55}$$

where b is the integration constant. By taking into account the geometrical relation $c_p(x) = \frac{\sin \psi(x)}{x}$, one gets the equation

$$\sin \psi(x) = \mathrm{h}x + \frac{\mathrm{b}}{x}. \tag{6.56}$$

The last equation is the exact Gauss map of the surface. From the two obvious geometrical conditions

$$\mathrm{h}a + \frac{\mathrm{b}}{a} = 1, \qquad \mathrm{h}c + \frac{\mathrm{b}}{c} = 0, \tag{6.57}$$

we can find one-to-one relations between physical h, b and geometrical a, c parameters, i.e.,

$$\mathrm{h} = \frac{a}{a^2 - c^2}, \qquad \mathrm{b} = -\frac{ac^2}{a^2 - c^2}. \tag{6.58}$$

Finally, the integration of the slope equation

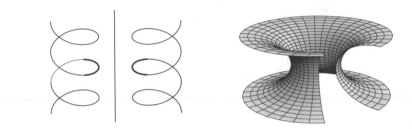

Fig. 6.14 The profile curves of the nodoid (*left*, solid parts) generating the stalk surface under revolution and a 3D view of the open part of the stalk (*right*)

$$\frac{dz}{dx} = \tan \psi(x) \tag{6.59}$$

gives us the profile curve

$$z = \int \tan \psi(x) dx = a \int_a^x \frac{(x^2 - c^2) dx}{\sqrt{(x^2 - a^2)(c^4 - a^2 x^2)}}. \tag{6.60}$$

The parameterization of the contour can be done using the elliptic functions, i.e.,

$$x(u) = \frac{c^2}{a} \mathrm{dn}(u, k), \qquad k = \frac{\sqrt{c^4 - a^4}}{c^2} \tag{6.61}$$

$$z(u) = \frac{c^2}{a} E(\mathrm{am}(u, k), k) - a F(\mathrm{am}(u, k), k).$$

Plots of the meridional sections of the cell fusion resulting from those expressions are shown in Fig. 6.14 (left), and a 3D view of the stalk is pictured on the right.

6.3.3   Geometric and Energetic Aspects

Having the explicit parameterization (6.61) of the profile curves, it is a simple matter to write down the parameterization of the relevant part of the stalk surfaces in the form

$$\tilde{\mathbf{x}} = (x(u) \cos v, x(u) \sin v, z(u) - z(K(k))), \qquad v \in (0, 2\pi). \tag{6.62}$$

With this at hand, it is also easy to find the coefficients E, F, and G of the first fundamental form of the surface (6.62)

Fig. 6.15 First curve is produced via the analytical result (6.65), and the second one is plotted by making use of the approximation formula presented in Markin and Albanesi (2002).

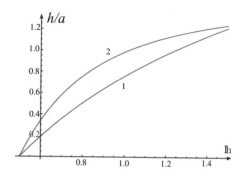

$$E = \tilde{\mathbf{x}}_u.\tilde{\mathbf{x}}_u = \frac{(a^2 - c^2)^2 \mathrm{dn}^2(u, k)}{a^2}$$

$$F = \tilde{\mathbf{x}}_u.\tilde{\mathbf{x}}_v = 0 \qquad\qquad (6.63)$$

$$G = \tilde{\mathbf{x}}_v.\tilde{\mathbf{x}}_v = \frac{c^4 \mathrm{dn}^2(u, k)}{a^2}.$$

These coefficients then give the infinitesimal element $\mathrm{d}\mathcal{A} = \sqrt{EG - F^2}\,\mathrm{d}u \wedge \mathrm{d}v$ of the surface area

$$\mathrm{d}\mathcal{A} = \frac{c^2(c^2 - a^2)\mathrm{dn}^2(u, k)}{a^2}\mathrm{d}u \wedge \mathrm{d}v, \qquad u \in \left\{\frac{K(k)}{2}, \frac{3K(k)}{2}\right\}. \qquad (6.64)$$

Integrating over the whole surface of the neck, one easily finds that the energy of the stalk is given by the formula

$$W_s = \frac{1}{2}k\left[(c_{\mathrm{stalk}} - \mathbb{h})^2 - \mathbb{h}^2\right]A = 2\pi k\frac{c^2(c^2 - a^2)}{a^2}E(t, k)|_{K(k)/2}^{K(k)}.$$

With the explicit parameterization (6.62) of the stalk, one can also find its height $2h$ (see Fig. 6.13) by the formula

$$h = z(K(k)/2) - z(K(k)). \qquad (6.65)$$

This allows us to plot the dimensionless normalized distance h/a, which depends only on a single parameter c (i.e., $\mathbb{h}$), and to compare it with the approximate function presented by Markin and Albanesi (2002). The result is depicted as curve 1 in Fig. 6.15.

It should be noted that the resulting plot in Fig. 6.15 has an universal character which is applicable to any spontaneous curvature.

Comments

Fusion involves drastic, although local, changes in the initial membrane structure. The membrane configurations emerging at the intermediate stages of fusion require inputs of energy and, hence, represent energy barriers the membranes have to overcome on the way to the new fused state. Those energy barriers are essential determinants of the fusion rate. The free energy of fusion stalks has been calculated through different approaches.

For example, Kuzmin et al. (2001) suggested a theoretical model that includes, besides bending, a tilt of the lipid molecules. The model starts from preformed nipples that decrease the local distance of two fusing membranes and requires an extraordinarily high energy to form a stalk out of two apposed, planar bilayers. The geometry of this model, however, is predefined.

Markin and Albanesi (2002) postulated a stress free stalk. The key point of their model is the optimization of the cross-sectional shape of the stalk's neck in terms of its bending energy. Relying on numerics, they did not recognize that this is a constant mean curvature surface, which is the main point of the present study.

6.4 Cylindrical Membranes

6.4.1 Translational–Invariant Solutions

The methods in Sect. 5.5.2 are basic for the derivation and detailed study of a new class of analytical solutions to (5.46), which describe the cylindrical forms and provide an exhaustive list of all such solutions. The respective parameterizations are obtained in explicit form, and the conditions for closing these generalized cylindrical surfaces, as well as the necessary and sufficient condition for their intersections or non-intersections, are found.

For the given values of the parameters k_c, $\mathbb{h}$, λ and p, the function $v(x^1)$ satisfying Eq. (5.48) describes a translational-invariant solution of the membrane shape equation (5.46), which corresponds to a cylindrical surface in $\mathbb{R}^3$ with generating lines parallel to the axis OY whose directrices are plane curves Γ of curvature $\kappa(s) = 2H(s)$. For such surfaces, the shape equation (5.46) reduces to the single ordinary differential equation

$$2\frac{\mathrm{d}^2\kappa\,(s)}{\mathrm{d}s^2} + \kappa^3\,(s) - \mu\kappa\,(s) - \sigma = 0, \qquad (6.66)$$

where

$$\mu = \mathbb{h}^2 + \frac{2\lambda}{k_c}, \qquad \sigma = -\frac{2p}{k_c}.$$

Fig. 6.16 Intersection of the infinite generalized cylinder with the plane $Y \equiv 0$. The slope angle is $\varphi(s)$, the angle between the position vector $\mathbf{x}(0)$ and $\mathbf{x}(s)$ is $\theta(s)$, and $\mathbf{t}(s)$ - tangent vector at the point $\mathbf{x}(s)$

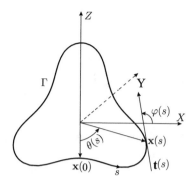

We will call (6.66) the equation of the generalised elasticas. If one of its solutions is known explicitly for $\sigma \neq 0$, the parametric equation of the respective curve Γ has the form (up to translation in the plane $\mathbb{R}^2$)

$$x(s) = \frac{2}{\sigma} \frac{d\kappa(s)}{ds} \cos \varphi(s) + \frac{1}{\sigma}(\kappa^2(s) - \mu) \sin \varphi(s)$$
$$z(s) = \frac{2}{\sigma} \frac{d\kappa(s)}{ds} \sin \varphi(s) - \frac{1}{\sigma}(\kappa^2(s) - \mu) \cos \varphi(s), \tag{6.67}$$

where (see Fig. 6.16) $\varphi(s)$ is the angle between the tangent vector $\mathbf{t}(s)$ of the curve Γ and the OX axis (slope angle), and thus is given by the formula

$$\varphi(s) = \int \kappa(s)\, ds. \tag{6.68}$$

For $\sigma = 0$, the situation is quite similar, namely,

$$x(s) = \frac{1}{\kappa^2(0) - \mu} \int \kappa^2(s)\, ds - \frac{\mu s}{\kappa^2(0) - \mu}, \quad z(s) = -\frac{2\kappa(s)}{\kappa^2(0) - \mu}. \tag{6.69}$$

The main goal is to find all solutions to the membrane shape equation (6.66) and after that to solve the integral (6.68) or the integral in parametric equation (6.69).

6.4.2 Analytical Solutions

In Vassilev et al. (2006, 2007), it was shown that every solution $\kappa(s)$ to Eq. (6.66) with coefficients σ and μ obeys a conservation law of type

$$\dot{\kappa}^2 = P(\kappa), \quad P(\kappa) = 2E - \frac{1}{4}\kappa^4 + \frac{1}{2}\mu\kappa^2 + \sigma\kappa, \tag{6.70}$$

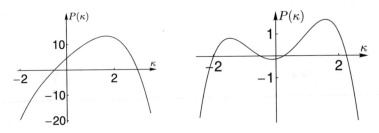

Fig. 6.17 Polynomial $P(\kappa)$: *left* - Case (I), *right* - Case (II)

where E is a real number that is the total energy of the solution. Thus, the general claim is that only two alternative possibilities have to be considered, namely: (I) the polynomial $P(\kappa)$ has two simple real roots $\alpha, \beta \in \mathbb{R}$, $\alpha < \beta$ and a pair of complex conjugate roots $\gamma, \delta \in \mathbb{C}$, $\delta = \bar{\gamma}$, (II) the polynomial $P(\kappa)$ has four simple real roots $\alpha < \beta < \gamma < \delta \in \mathbb{R}$ (see Fig. 6.17).

The explicit analytic expressions for the solutions to Eq. (6.66), and the corresponding slope angles are given by Theorem 6.1 for Case (I), and in Theorem 6.2 for Case (II).

Theorem 6.1 (Vassilev et al. (2008)). *Given μ, σ and E, let the roots α, β, γ and δ of the respective polynomial $P(\kappa)$ be as described in Case (I), that is, $\alpha < \beta \in \mathbb{R}$, $\gamma, \delta \in \mathbb{C}$, $\delta = \bar{\gamma}$. Let $\eta \equiv (\gamma - \bar{\gamma})/2i$. Then, except for the cases in which $(3\alpha + \beta)(\alpha + 3\beta) = \eta = 0$, the function*

$$\kappa_1(s) = \frac{(A\beta + B\alpha) - (A\beta - B\alpha)\operatorname{cn}(us, k)}{(A + B) - (A - B)\operatorname{cn}(us, k)} \tag{6.71}$$

of the real variable s, where

$$A = \sqrt{4\eta^2 + (3\alpha + \beta)^2}, \quad B = \sqrt{4\eta^2 + (\alpha + 3\beta)^2}, \quad u = \frac{1}{4}\sqrt{AB} \tag{6.72}$$

$$k = \frac{1}{\sqrt{2}}\sqrt{1 - \frac{4\eta^2 + (3\alpha + \beta)(\alpha + 3\beta)}{\sqrt{[4\eta^2 + (3\alpha + \beta)(\alpha + 3\beta)]^2 + 16\eta^2(\beta - \alpha)^2}}}, \tag{6.73}$$

is the curvature which takes real values for each $s \in \mathbb{R}$ and satisfies Eq. (6.66). This function is periodic if and only if $\eta \neq 0$ or $\eta = 0$, but if $(3\alpha + \beta)(\alpha + 3\beta) > 0$, its least period is $T_1 = (4/u) K(k)$, and for $\sigma \neq 0$, its indefinite integral $\varphi_1(s)$, such that $\varphi_1(0) = 0$, is

$$\varphi_1(s) = \frac{A\beta - B\alpha}{A - B}s + \frac{(A + B)(\alpha - \beta)}{2u(A - B)}\Pi\left(-\frac{(A - B)^2}{4AB}, \operatorname{am}(us, k), k\right)$$

$$+ \frac{\alpha - \beta}{2u\sqrt{k^2 + \frac{(A-B)^2}{4AB}}}\arctan\left(\sqrt{k^2 + \frac{(A - B)^2}{4AB}}\frac{\operatorname{sn}(us, k)}{\operatorname{dn}(us, k)}\right). \tag{6.74}$$

In the case in which $(3\alpha + \beta)(\alpha + 3\beta) = \eta = 0$, *the function*

$$\kappa_2(s) = \zeta - \frac{4\zeta}{1 + \zeta^2 s^2}, \tag{6.75}$$

where $\zeta = \alpha$ *when* $3\alpha + \beta = 0$ *and* $\zeta = \beta$ *when* $\alpha + 3\beta = 0$, *satisfies Eq.* (6.66), *for each* $s \in \mathbb{R}$, *and its indefinite integral* $\varphi_2(s)$ *for which* $\varphi_2(0) = 0$ *reads as*

$$\varphi_2(s) = \zeta s - 4\arctan(\zeta s). \tag{6.76}$$

Theorem 6.2 (Vassilev et al. (2008)). *Let* μ, σ *and* E *be given and let the respective roots* α, β, γ *and* δ *of the polynomial* $P(\kappa)$ *be as in Case* (II), *that is,* $\alpha < \beta < \gamma < \delta \in \mathbb{R}$. *Consider the functions*

$$\kappa_3(s) = \delta - \frac{(\delta - \alpha)(\delta - \beta)}{(\delta - \beta) + (\beta - \alpha)\operatorname{sn}^2(us, k)} \tag{6.77}$$

$$\kappa_4(s) = \beta + \frac{(\gamma - \beta)(\delta - \beta)}{(\delta - \beta) - (\delta - \gamma)\operatorname{sn}^2(us, k)} \tag{6.78}$$

of the real variable s, *where*

$$u = \frac{1}{4}\sqrt{(\gamma - \alpha)(\delta - \beta)}, \qquad k = \sqrt{\frac{(\beta - \alpha)(\delta - \gamma)}{(\gamma - \alpha)(\delta - \beta)}}. \tag{6.79}$$

It is clear that they take real values for each $s \in \mathbb{R}$, *and satisfy Eq.* (6.66). *Besides, they are periodic with least period* $T_2 = (2/u)K(k)$ *and the indefinite integrals* $\varphi_3(s)$ *and* $\varphi_4(s)$, *such that* $\varphi_3(0) = \varphi_4(0) = 0$ *are*

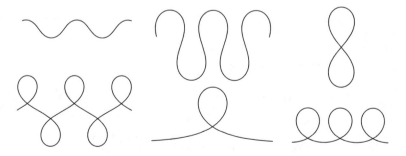

Fig. 6.18 Euler's elastica generated via formulas (6.69), and solutions to Eq. (6.66) with $\sigma = 0$

$$\varphi_3\,(s) = \delta s - \frac{\delta - \alpha}{u} \Pi \left(\frac{\beta - \alpha}{\beta - \delta}, \, \mathrm{am}(us, k), k \right) \tag{6.80}$$

$$\varphi_4\,(s) = \beta s - \frac{\beta - \gamma}{u} \Pi \left(\frac{\delta - \gamma}{\delta - \beta}, \, \mathrm{am}(us, k), k \right). \tag{6.81}$$

In this way, we have an analytical description of all cylindrical surfaces whose curvatures satisfy Eq. (6.66).

Remark 6.1 Formulae (6.67) and (6.70) imply the remarkable relation

$$\kappa = \frac{\sigma}{4} r^2 - \frac{8E + \mu^2}{4\sigma}, \tag{6.82}$$

which tell us that the curvatures of the generalized elasticas defined by Eq. (6.66) are simply functions of the polar radius r. In the reverse direction, if we take as a curvature the expression (6.82), it can be proven that the intrinsic equation of the respective curve is exactly (6.66).

6.4.3   Closure Conditions

Hereafter, we are interested in directrices Γ, which close up smoothly, meaning that there exists a value L of the arclength s such that $\mathbf{x}(0) = \mathbf{x}(L)$ and $\mathbf{t}(0) = \mathbf{t}(L)$. The latter property of such a smooth closed directrix Γ and the definition of the tangent vector imply that there exists an integer m such that $\varphi(L) = \varphi(0) + 2m\pi$. Since in the above theorems, it was chosen that $\varphi(0) = 0$, this means that the condition $\mathbf{t}(0) = \mathbf{t}(L)$ leads to the relation

$$\varphi\,(T) = \frac{2m\pi}{n}. \tag{6.83}$$

The last relation is clearly the necessary condition for the closure of the directrix Γ. Apparently, it is a sufficient condition as well.

Straightforward computations lead to the following explicit expressions for the angles in formulas (6.80) and (6.81):

$$\varphi_1\,(T_1) = \frac{4\,(A\beta - B\alpha)}{u\,(A - B)} K\,(k) + 2\frac{(A + B)\,(\alpha - \beta)}{u\,(A - B)} \Pi \left(-\frac{(A - B)^2}{4AB}, k \right) \tag{6.84}$$

$$\varphi_3\,(T_2) = \frac{2\delta}{u} K\,(k) + 2\frac{\alpha - \delta}{u} \Pi \left(\frac{\alpha - \beta}{\delta - \beta}, k \right) \tag{6.85}$$

$$\varphi_4\,(T_2) = \frac{2\beta}{u} K\,(k) + 2\frac{\gamma - \beta}{u} \Pi \left(\frac{\gamma - \delta}{\beta - \delta}, k \right). \tag{6.86}$$

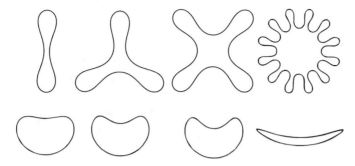

Fig. 6.19 Examples of closed directrices without self-intersections, drawn in the fixed (*up*) and the moving frame (*down*)

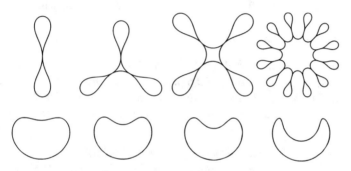

Fig. 6.20 Examples of closed directrices with an internal contact drawn in the fixed (*up*) and the moving frame (*down*)

These expressions and the closure condition (6.83) allows us to determine whether a directrix Γ, whose curvature is either (6.71), (6.77) or (6.78), closes up smoothly or not.

An interesting property of the curves with curvatures $\kappa_3(s)$ and $\kappa_4(s)$, which is related to the closure condition (6.83), is that if one of these curves closes up smoothly, then the other one does as well.

6.4.4 Self-Intersections

In what concerns the vesicle shapes, of special interest are solutions to Eq. (6.66) that give rise to closed non-self-intersecting (simple) curves. A sufficient condition for such a closed curve to be simple is $\mu < 0$, which is discussed in Capovilla et al. (2002). It is also mentioned therein that the closed curves satisfying condition (6.83) with $m \neq \pm 1$ or $n = 1$ are necessarily self-intersecting. The case of $\mu > 0$ is considered, and several new sufficient conditions are presented for a closed curve of

Fig. 6.21 Examples of closed directrices with self-intersections, drawn in the fixed (*up*) and the moving frame (*down*)

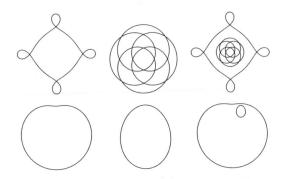

the foregoing type to satisfy a closure condition of form (6.83) with $m = \pm 1$ and $n \geq 2$.

The main result is the following:

Theorem 6.3 (Djondjorov et al. (2012)). *Let Γ be a smooth closed curve, whose curvature $\kappa(s)$ is a solution to Eq. (6.66) in one of the forms which appear in (6.71), (6.77) or (6.78). Let T be the least period of the function $\kappa(s)$ and let the angle $\varphi(s)$, corresponding to $\kappa(s)$ according to formula (6.68), satisfy a closure condition of form (6.83) with $m = \pm 1$ and $n \geq 2$. Then,*

(i)   the curve Γ is simple if $\kappa^2(s) - \mu \neq 0$ for $s \in [0, T/2]$
(ii) the curve Γ is self-intersecting if the equation $\kappa^2(s) - \mu = 0$ has exactly one solution for $s \in [0, T/2]$.

Below, we present another immediate consequence of the above theorem (Fig. 6.21).

Corollary 6.1 *Under the assumptions of Theorem 6.3, the curve Γ is self-intersecting if its curvature $\kappa(s)$ is such that the polynomial $P(\kappa)$ has only real roots.*

6.4.5   Hele-Shaw Cells

Most readers probably think that our considerations in the last two sections are dictated by a purely academic interest. The principal aim of the present section is to convince them that these studies are related to some real physical phenomena that have a lot of practical applications.

We start by recalling that in the literature, it is well known that the loos of stability of the interface between two fluids (the so-called Saffman and Taylor (1958) instability) is due to the formation of patterns of curvature in this interface, which differs significantly from the curvature in other parts of the fluid boundary. This phenomena occurs, e.g., when the Hele-Shaw cell contains fluid into which less viscous fluid has been injected (see dos Reis and Miranda (2011)). It can also occur due to gravity

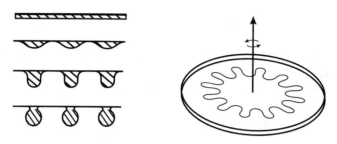

Fig. 6.22 Hele-Shaw cell—vertical (*left*) and rotating with a constant angular speed (*right*)

if a horizontal interface separates two fluids of different densities and the heavier fluid is above the other one, as shown by Bratsun and DeWit (2011) and Schwarzenberger et al. (2012) (see Fig. 6.22, left). This type of instability is characteristic of a Hele-Shaw cell subjected to pressure, a magnetic field or a rotation (Fig. 6.22, right).

Nye et al. (1984) have experimentally considered the type of instability that is due to gravitation (Fig. 6.22, left), using oil on the top and air underneath, and by relying on hydrodynamic arguments about the curvature κ, they have concluded that

$$\kappa \sim z.$$

According to the results obtained earlier (cf. 4.97), this curvature type generates the Euler elasticas presented in Fig. 6.18.

Other authors, such as Leandro et al. (2008) and Oliveira et al. (2008), have shown that in a rotating Hele-Shaw cell, the equation describing the balance of centrifugal forces and surface tension can be integrated and the curvature between the two fluids can be expressed as

$$\kappa(r) = \Omega(r^2 - \mathring{r}^2), \tag{6.87}$$

where $r = \sqrt{x^2 + z^2}$ is the radius, Ω is the non-dimensional angle velocity, and $\mathring{r}$ is the radius for which the curvature $\kappa(r)$ became zero.

It is not difficult to realize that, following the scheme for the reconstruction of curves presented in Sect. 2.2, the curvature in (6.87) satisfies an intrinsic equation of type (6.70), and therefore the results from Sects. 6.4.2-6.4.4 are valid (up to appropriate changes of the coefficients).

In Fig. 6.19, examples are shown of curves, generated using solutions described in Sect. 6.4.3 that are appropriate for the observed interface shapes (Djondjorov et al. (2012a)).

In a rotating Hele-Shaw cell, there exist values of the angular velocity at which the interface exhibits shapes with points of contact that can be considered as the onset of drop formation and separation (see Fig. 6.20).

6.5 Beyond Delaunay Surfaces

Many years ago, Kenmotsu (1980) showed that surfaces of a given mean curvature in $\mathbb{R}^3$ are defined essentially by their Gauss map. Later on, Eells (1987) pointed out that the Gauss map for Delaunay surfaces is given by the formula

$$\sin\theta = ax + \frac{c}{x}, \qquad x \neq 0, \qquad a, c \in \mathbb{R}. \tag{6.88}$$

Finally, in 1995, Naito et al. (1995) discovered (see also Ou-Yang et al. (1999)) that (6.88), which is a solution to the reduced equation

$$\cos^3\theta \frac{d^3\theta}{dx^3} = 4\sin\theta\cos^2\theta \frac{d^2\theta}{dx^2}\frac{d\theta}{dx} - \cos\theta\left(\sin^2\theta - \frac{1}{2}\cos^2\theta\right)\left(\frac{d\theta}{dx}\right)^3$$

$$+ \frac{7\sin\theta\cos^2\theta}{2x}\left(\frac{d\theta}{dx}\right)^2 - \frac{2\cos^3\theta}{x}\frac{d^2\theta}{dx^2} \tag{6.89}$$

$$+ \left(\frac{\lambda}{k_c} + \frac{\mathrm{h}^2}{2} - \frac{2\mathrm{h}\sin\theta}{x} - \frac{\sin^2\theta - 2\cos^2\theta}{2x^2}\right)\cos\theta\frac{d\theta}{dx}$$

$$+ \left(\frac{\lambda}{k_c} + \frac{\mathrm{h}^2}{2} - \frac{\sin^2\theta + 2\cos^2\theta}{2x^2}\right)\frac{\sin\theta}{x} + \frac{p}{k_c}$$

of the shape equation (5.46) describing axially symmetric constant mean curvature surfaces, could be generalized to the form

$$\sin\theta = \varepsilon + \frac{1}{\mathrm{h}x} + \frac{1}{4}\left(\varepsilon^2 + 2\right)\mathrm{h}x, \qquad \varepsilon \in \mathbb{R}, \tag{6.90}$$

which corresponds to vesicles with spontaneous curvature ($\mathrm{h} \neq 0$), subjected to nonzero pressure ($p \neq 0$), and provided that the pressure p and the tensile stress λ are given by the expressions

$$\frac{\lambda}{k_c} = \frac{\mathrm{h}^2\left(\varepsilon^2 + 1\right)}{2}, \qquad \frac{p}{k_c} = -\frac{\mathrm{h}^3\left(\varepsilon^2 + 2\right)^2}{8}.$$

According to Fig. 1.1 for the foregoing class of solutions defined by Eq. (6.90), we have

$$\frac{dz}{dx} = \frac{\varepsilon + \frac{1}{\mathrm{h}x} + \frac{1}{4}\mathrm{h}\left(\varepsilon^2 + 2\right)x}{\sqrt{1 - \left(\varepsilon + \frac{1}{\mathrm{h}x} + \frac{1}{4}\mathrm{h}\left(\varepsilon^2 + 2\right)x\right)^2}}, \tag{6.91}$$

and hence the profile curve of such an axisymmetric vesicle can be expressed as the graph $(x, z(x))$ of the function $z(x)$ given by the following elliptic integral

$$z(x) = \int \frac{\varepsilon + \frac{1}{\ln x} + \frac{1}{4}\left(\varepsilon^2 + 2\right)\ln x}{\sqrt{1 - \left(\varepsilon + \frac{1}{\ln x} + \frac{1}{4}\left(\varepsilon^2 + 2\right)\ln x\right)^2}}\, dx.$$

The principal goal of the next section is to find out parameterizations of the above-mentioned contours that are free of the obvious limitations associated with the graph presentations $(x, z(x))$ of the function $z(x)$.

6.5.1 Parametric Equations

In terms of an appropriate new variable u, Eq. (6.91) can be rewritten in the form

$$\frac{dx}{du} = \frac{1}{\mu}\sqrt{-P(x)Q(x)} \tag{6.92}$$

$$\frac{dz}{du} = \frac{1}{2\mu}\left(P(x) + Q(x)\right), \tag{6.93}$$

in which

$$P(x) = x^2 + \frac{4\left(\varepsilon - 1\right)}{\left(\varepsilon^2 + 2\right)\ln}x + \frac{4}{\left(\varepsilon^2 + 2\right)\ln^2} \tag{6.94}$$

$$Q(x) = x^2 + \frac{4\left(\varepsilon + 1\right)}{\left(\varepsilon^2 + 2\right)\ln}x + \frac{4}{\left(\varepsilon^2 + 2\right)\ln^2}, \tag{6.95}$$

and where the real parameter μ will be fixed later on.

It should be noticed that the roots of the polynomial $\Pi(x) = P(x)Q(x)$ are

$$\alpha = \frac{2\left(1 - \varepsilon - \sqrt{-2\varepsilon - 1}\right)}{\left(\varepsilon^2 + 2\right)\ln}, \qquad \beta = \frac{2\left(1 - \varepsilon + \sqrt{-2\varepsilon - 1}\right)}{\left(\varepsilon^2 + 2\right)\ln}$$

$$\gamma = \frac{2\left(-1 - \varepsilon + \sqrt{2\varepsilon - 1}\right)}{\left(\varepsilon^2 + 2\right)\ln}, \qquad \delta = \frac{2\left(-1 - \varepsilon - \sqrt{2\varepsilon - 1}\right)}{\left(\varepsilon^2 + 2\right)\ln},$$

and therefore, for each allowable value of the parameter ε, i.e., $|\varepsilon| > 1/2$, only two of them are real. These are α and $\beta \neq \alpha$ for $\varepsilon < -1/2$, and alternatively γ and $\delta \neq \gamma$ for $\varepsilon > 1/2$. In the first case, we will have $0 < \alpha \leq x \leq \beta$ when $\ln > 0$, and in the second case, x will be strictly positive, i.e., $0 < \gamma \leq x \leq \delta$ if and only if $\ln < 0$.

Now, using the standard procedure for handling elliptic integrals (see Whittaker and Watson (1922, Sect. 22.7)), one can express the solution $x(u)$ to Eq. (6.92) in the form

$$x(u) = \frac{2\,\mathrm{sign}\,(\varepsilon)}{\ln\sqrt{\varepsilon^2 + 2}}\left(1 - \frac{2\tau}{\tau + \mathrm{cn}(u, k)}\right), \tag{6.96}$$

where

$$\tau = \sqrt{\frac{1 + |\varepsilon| + \sqrt{2 + \varepsilon^2}}{1 + |\varepsilon| - \sqrt{2 + \varepsilon^2}}}, \qquad k = \sqrt{\frac{1}{2} - \frac{3}{4\sqrt{2 + \varepsilon^2}}} \cdot$$

Actually, the choice of u as a uniformization variable also fixes the value of the free parameter μ, i.e.,

$$\mu = \frac{4}{\hbar \left(2 + \varepsilon^2\right)^{3/4}} \cdot$$

Consequently, using expressions (6.94) and (6.95), one can write down the solution $z(u)$ of Eq. (6.93) in the form

$$z(u) = \frac{1}{\mu} \int \left(x^2(u) + \frac{4 \varepsilon x(u)}{\left(\varepsilon^2 + 2\right) \hbar} + \frac{4}{\left(\varepsilon^2 + 2\right) \hbar^2} \right) du, \qquad (6.97)$$

and following this route, we have found that

$$z(u) = \mu \left[E(\mathrm{am}(u, k), k) - \frac{\mathrm{sn}(u, k) \, \mathrm{dn}(u, k)}{\tau + \mathrm{cn}(u, k)} - \frac{u}{2} \right]. \qquad (6.98)$$

The meaning of the functions that appear in the above equation is as follows. $E(\cdot, \cdot)$ denotes the incomplete elliptic integral of the second kind, which depends on its argument and the *elliptic modulus* k. The Jacobian amplitude function $\mathrm{am}(\cdot, \cdot)$ and Jacobian elliptic functions $\mathrm{sn}(\cdot, \cdot)$, $\mathrm{cn}(\cdot, \cdot)$ and $\mathrm{dn}(\cdot, \cdot)$ are dependent in the same manner (see Appendix A.1).

In what follows, we will present an alternative parameterization of Delaunay-like surfaces which we hope will be of some help in their study from the geometrical viewpoint.

We start by rewriting (6.96) in the form (Djondjorov et al. (2010))

$$x(u) = -\frac{2\mathrm{sign}(\varepsilon) \left(1 + |\varepsilon| + \sqrt{2|\varepsilon| - 1}\right) \mathrm{dn} \left(\tilde{u}, m\right)}{\left(\varepsilon^2 + 2\right) \hbar}, \qquad (6.99)$$

where

$$\tilde{u} = \frac{K(m)}{2K(k)} u + K(m), \qquad m = \frac{2\sqrt{(1 + |\varepsilon|)}\sqrt{2|\varepsilon| - 1}}{1 + |\varepsilon| + \sqrt{2|\varepsilon| - 1}}, \qquad (6.100)$$

and $K(\cdot)$ denotes the complete elliptic integral of the first kind evaluated for the respective elliptic modulus.

If one takes into account that we have the formulas

$$\int \mathrm{dn}(t, k) dt = \mathrm{am}(t, k), \qquad \int \mathrm{dn}^2(t, k) dt = E(\mathrm{am}(t, k), k), \qquad (6.101)$$

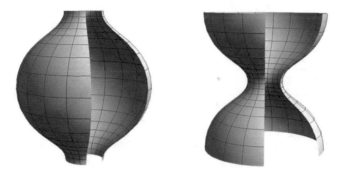

Fig. 6.23 Open parts of the bulb (*left*) and the neck (*right*) segments of the periodic surface of revolution obtained via parametric equations (6.99) and (6.103) with $\varepsilon = 1.3542$ and $\mathbb{h} = -3.335623$

the remaining integrations in (6.97) are straightforward.

Actually, the integration produces the primitive

$$\zeta(u) = \frac{8K(k)}{\mu\mathbb{h}^2(\varepsilon^2+2)K(m)} \left(\frac{(1+|\varepsilon|+\sqrt{2|\varepsilon|-1})^2}{\varepsilon^2+2} E(\mathrm{am}(\tilde{u},m),m) \right.$$
$$\left. + \frac{2\mathrm{sign}(\mathbb{h})\varepsilon(1+|\varepsilon|+\sqrt{2|\varepsilon|-1})}{\varepsilon^2+2} \mathrm{am}(\tilde{u},m) + F(\mathrm{am}(\tilde{u},m),m) \right), \tag{6.102}$$

in which the integration constant is omitted because if we want the sought-after curve to start from the OX axis for $u = 0$, then obviously, for the ordinate $z(u)$, we have to take

$$z(u) = \zeta(u) - \zeta(0). \tag{6.103}$$

Thus, for each pair of the allowed values of the parameters ε and $\mathbb{h}$, the expressions (6.99) and (6.103) provide the parametric equations of the profile curves of our axially symmetric unduloid-like surfaces corresponding to the respective solutions of the membrane shape equation (6.89) of the form (6.90) (see Fig. 6.23).

We will make the following comments. The first one is that if we equate the right-hand sides of, respectively, Eqs. (6.96) and (6.99) and Eqs. (6.98) and (6.103), we will face quite nontrivial relationships among elliptic functions and integrals. It is hard to believe that they could be derived in a purely analytic way and probably should simply be considered as glimpses of the underlying geometry.

The second one concerns the studies of the surfaces of revolution with periodic mean curvature undertaken by Kenmotsu (2003), who had presented numerical examples of such surfaces. To our knowledge, the surfaces presented here provide the first examples from this class in analytical form.

References

B. Allen, S. Rogers, J. Ghilardi, P. Menning, M. Kuskovsky, I. Baubaum, D. Simone, P. Manyth, Noxious cutaneous thermal stimuli induce a graded release of endogenous substance P in the spinal cord: Imaging peptide action in vivo. J. Neurosci **17**, 5221–5227 (1997)

R. Bar-Ziv, E. Moses, Instability and pearling states produced in tubular membranes by competition of curvature and tension phys. Rev. Lett. **73**, 1392–1395 (1994)

D. Bratsun, A. DeWit, Buoyancy-driven pattern formation in reactive immiscible two-layer systems. Chem. Eng. Sci. **66**, 5723–5734 (2011)

P. Byrd, M. Friedman, *Handbook of Elliptic Integrals for Engineers and Scientists*, 2nd edn. (Springer, Berlin, 1971)

R. Capovilla, C. Chryssomalakos, J. Guven, Elastica hypoarealis. Eur. Phys. J. B **29**, 163–166 (2002)

K. Cole, Surface forces of the arbacia egg. J. Cell. Comp. Physiol. **1**, 1–9 (1932)

C. Delaunay, Sur la surface de revolution dont la courbure moyenne est constante. J. Math. Pures et Appliquées **6**, 309–320 (1841)

P. Djondjorov, M. Hadzhilazova, I. Mladenov, V. Vassilev, Beyond Delaunay surfaces. J. Geom. Symmetry Phys. **18**, 1–11 (2010)

P. Djondjorov, V. Vassilev, I. Mladenov, Deformation of injected vesicles adhering onto flat rigid substrates. Comput. Math. Appl. **64**, 214–220 (2012)

P. Djondjorov, V. Vassilev, M. Hadzhilazova, I. Mladenov, Analytic description of the viscous fingering interface in a Hele-Shaw cell. Geom. Integr. Quant. **13**, 107–113 (2012a)

L. dos Reis, A. Miranda, Controlling fingering instabilities in nonflat Hele-Shaw geometries. Phys. Rev. E **84**, 066313 (2011)

J. Eells, The surfaces of Delaunay. Math. Intell. **9**, 53–57 (1987)

M. Hadzhilazova, J.-F. Ganghoffer, I. Mladenov, Analytical description of the shapes of beaded nerve fibres. CRAS (Sofia) **63**, 1155–1162 (2010)

M. Hadzhilazova, I. Mladenov, Surface tension via Cole's experiment, in *Proceedings of the Tenth International Summer School of Chemical Engineering*, (Sofia, 2004) pp. 195–200

M. Hadzhilazova, I. Mladenov, A mathematical examination of squeezing and stretching of spherical vesicles. Geom. Integr. Quant. **6**, 231–239 (2005)

M. Hadzhilazova, I. Mladenov, J. Oprea, Unduloids and their geometry. Archivum Mathematicum **43**, 417–429 (2007)

K. Kenmotsu, Surfaces of revolution with prescribed mean curvature. Tohoku Math. J. **32**, 147–153 (1980)

K. Kenmotsu, Surfaces of revolution with periodic mean curvature. Osaka J. Math. **40**, 687–696 (2003)

M. Kozlov, V. Markin, Possible mechanism of membrane fusion. Biofizika **28**, 242–247 (1983)

P. Kuzmin, J. Zimmerberg, Y. Chizmadzhev, F. Cohen, A quantitative model for membrane fusion based on low-energy intermediates. Proc. Natl. Acad. Sci. **98**, 7235–7240 (2001)

E. Leandro, R. Oliveira, J. Miranda, Geometric approach to stationary shapes in rotating Hele-Shaw flows. Phys. D Nonlinear Phenom. **237**, 652–664 (2008)

P. Manthy, C. Allen, J. Ghilardi, Rapid endocytosis of a G protein-coupled receptor: Substance P-Evoked internalization of its receptor in the rats triatum in vivo. Proc. Natl. Acad. Sci. USA **92**, 2622–2626 (1995)

V. Markin, D. Tanellian, R. Jersild, S. Ochs, Biomechanics of stretch-induced beading. Biophys. J. **78**, 2852–2860 (1999)

V. Markin, J. Albanesi, Membrane fusion: Stalk model revisited. Biophys. J. **82**, 693–712 (2002)

I. Mladenov, Delaunay surfaces revisited. CRAS (Sofia) **55**, 19–24 (2002a)

I. Mladenov, New solutions of the shape equation. Eur. Phys. J. B **29**, 327–330 (2002)

H. Naito, M. Okuda, Z.-C. Ou-Yang, New solutions to the Helfrich variation problem for the shapes of lipid bilayer vesicles: Beyond Delaunay's surfaces. Phys. Rev. Lett. **74**, 4345–4348 (1995)

J. Nye, H. Lean, A. Wright, Interfaces and falling drops in a Hele-Shaw cell. Eur. J. Phys. **5**, 73–80 (1984)

S. Ochs, R. Jersild, Myelin intrusions in beaded nerve fibers. Neuroscience **36**, 553–567 (1990)

S. Ochs, R. Pourmand, R. Jersild Jr., Origin of beading constrictions at the axolemma: Presence in unmyelinated axons and after β, β'-Iminodipropionitrile(IDPN) degradation of the cytoskeleton. Neuroscience **70**, 1081–1096 (1996)

S. Ochs, R. Pourmand, R. Jersild Jr., R. Friedman, The origin and nature of beading: A reversible transformation of the shape of nerve fibers. Neurobiol **52**, 391–426 (1997)

R. Oliveira, J. Miranda, E. Leandro, Ferrofluid patterns in a radial magnetic field: Linear stability, nonlinear dynamics, and exact solutions. Neuroscience **77**, 016304 (2008)

J. Oprea, *Differential Geometry and Its Applications*, 3rd edn. (Mathematical Association of America, Washington D. C, 2007)

Z.-C. Ou-Yang, J.-X. Liu, Y.-Z. Xie, *Geometric Methods in the Elastic Theory of Membranes in Liquid Crystal Phases* (World Scientific, Hong Kong, 1999)

P. Saffman, G. Taylor, The penetration of a fluid into a porous medium or Hele-Shaw cell containing a more viscous fluid. Proc. Roy. Soc. A **245**, 312–329 (1958)

K. Schwarzenberger, K. Eckert, S. Odenbach, Relaxation oscillations between Marangoni cells and double diffusive fingers in a reactive liquid-liquid system. Chem. Eng. Sci. **68**, 530–540 (2012)

D. Tanelian, V. Markin, Biophysical and functional consequences of receptor-mediated nerve fiber transformation. Biophys. J. **72**, 1092–1108 (1997)

V. Vassilev, P. Djonjorov, I. Mladenov, Cylindrical equilibrium shapes of fluid membranes. J. Phys. A: Math. Theor. **41**, 435201–16 (2008)

V. Vassilev, P. Djonjorov, I. Mladenov, Symmetry groups, conservation laws and group-invariant solutions of the membrane shape equation. Geom. Integr. Quant. **7**, 265–279 (2006)

V. Vassilev, P. Djonjorov, I. Mladenov, On the translationally-invariant solutions of the membrane shape equation. Geom. Integr. Quant. **8**, 312–321 (2007)

E. Whittaker, G. Watson, *A Course of Modern Analysis*, (Cambridge University Press, Cambridge, 1922)

M. Yoneda, Tension at the surface of sea-urchin egg: A critical examination of Cole's experiment. J. Exp. Biol. **41**, 893–906 (1964)

H. Zhu, S. Wu, S. Schachter, Site-specific and sensory neuron-dependent increases in postsynaptic glutamate sensitivity accompany serotonin-induced long-term facilitation at aplysia sensomotor synapses. J. Neurosci. **17**, 4976–4985 (1997)

Epilogue

Everything created by humans has its beginning and its end. At the end of this book, we would like to say that this is just a first step towards a much deeper exposition of the topics discussed inside. First, because they are not treated with the completeness that we would like, and second, because there remain many examples in the physical world in which the elastica plays a principal role but which were not touched on here.

We concretely have in mind the free rotations of a rigid body about a fixed point, the Dubins problem concerning the flight of airplanes, trajectories of a material ball rolling in the plane, the three-dimensional elasticas, and probably many others which we do not suspect at this time.

We are, however, firmly convinced that glimpses of elastica are everywhere around us, and we have just to feel their presence and be struck again with admiration by this confluence of nature and geometry.

We will certainly be thankful to everyone who sends us comments, remarks or suggestions.

© Springer International Publishing AG 2017
I.M. Mladenov and M. Hadzhilazova, *The Many Faces of Elastica*,
Forum for Interdisciplinary Mathematics 3, DOI 10.1007/978-3-319-61244-7

Appendix A
Elliptical Integrals and Functions

Elliptic integrals and functions are mathematical objects, which nowadays are often omitted in the mathematical curricula of universities. One quite trivial explanation is the presence of plenty of efficient computational programs that can be implemented on modern computers. While the standard integration techniques allow us to obtain explicit expressions (in terms of trigonometric, exponential and logarithmic functions) for every integral of the form

$$\int \mathcal{R}(x, \sqrt{P(x)}) dx,$$

where $\mathcal{R}(x, \sqrt{P(x)})$ is a rational function, and $P(x)$ is a linear or quadratic polynomial, we have to widen our vocabulary of "elementary" functions if we want to work with polynomials of higher degree. In particular, when $P(x)$ is a polynomial of the third or fourth degree, the corresponding function is called **elliptic**. Of course, when teaching calculus, one must stop somewhere, and it is reasonable to stay loyal to well-known linear and quadratic functions, while using numerical methods to calculate integrals of the third and fourth degree. The possibilities of easy-to-use computer systems for symbolic manipulation of the type represented by *Maple*® and Mathematica® makes this course of action even more understandable.

The main point in this Appendix is that elliptic functions provide effective means for the description of geometric objects. The second is that the above-mentioned computer programs, through their built-in tools for calculation and visualization, are, in fact, a real motivation for the teaching and using of elliptic functions.

In this Appendix, we will consider a few examples in order to prove that elliptic integrals and functions are necessary to get more interesting geometric and mechanical information than that given by direct numerical calculations.

The history of the development of elliptic functions can be followed in Stillwell (1989). Clear statements of their properties and applications can be found in the books by Greenhill (1959), Hancock (1958), Bowman (1953) and Lawden (1989). A more recent approach to the problem from the viewpoint of dynamical systems is given by Meyer (2001).

© Springer International Publishing AG 2017
I.M. Mladenov and M. Hadzhilazova, *The Many Faces of Elastica*,
Forum for Interdisciplinary Mathematics 3, DOI 10.1007/978-3-319-61244-7

A.1 Jacobian Elliptic Functions

The easiest way to understand elliptic functions is to consider them as analogous to ordinary trigonometric functions. From the calculus, we know that

$$\arcsin(x) = \int_0^x \frac{du}{\sqrt{1 - u^2}}.$$

Of course, if $x = \sin(t)$, $-\pi/2 \le t \le \pi/2$, we will have

$$t = \arcsin(\sin(t)) = \int_0^{\sin(t)} \frac{du}{\sqrt{1 - u^2}}. \tag{A.1}$$

In this case, we can consider $\sin(t)$ as the inverse function of the integral (A.1). The real understanding of trigonometric functions includes knowledge of their graphs, their connection with other trigonometric functions, such as in $\sin^2(\theta) + \cos^2(\theta) = 1$, and of course, the fundamental geometric and physical parameters in which they are included (i.e., circumferences and periodical movements). We will follow this example for elliptic functions too.

Let us begin by fixing some k, $0 \le k \le 1$, which, from now on, will be called an **elliptic modulus** and introduce the following:

Definition A.1 The Jacobi sine function $\mathrm{sn}(u, k)$ is the inverse function of the following integral:

$$u = \int_0^{\mathrm{sn}(u,k)} \frac{dt}{\sqrt{1 - t^2}\sqrt{1 - k^2 t^2}}. \tag{A.2}$$

More generally, we will call

$$F(z, k) = \int_0^z \frac{dt}{\sqrt{1 - t^2}\sqrt{1 - k^2 t^2}} \tag{A.3}$$

the elliptic integral of the first kind. The elliptic integrals of the second and third kinds are defined by the equations

$$E(z, k) = \int_0^z \frac{\sqrt{1 - k^2 t^2}}{\sqrt{1 - t^2}} \, dt \tag{A.4}$$

$$\Pi(n, z, k) = \int_0^z \frac{dt}{(1 + nt^2)\sqrt{(1 - t^2)(1 - k^2 t)}}.$$

When the argument z in $F(z, k)$, $E(z, k)$ and $\Pi(n, z, k)$ is equal to one, these integrals are denoted, respectively, as $K(k)$, $E(k)$ and $\Pi(n, k)$ and called complete elliptic integrals of the first, second and third kinds, respectively.

If we put $t = \sin \phi$, the above integrals are transformed, respectively, into

$$F(\phi, k) = \int_0^\phi \frac{d\phi}{\sqrt{1 - k^2 \sin^2 \phi}}$$

$$E(\phi, k) = \int_0^\phi \sqrt{1 - k^2 \sin^2 \phi}\, d\phi \qquad (A.5)$$

$$\Pi(n, \phi, k) = \int_0^\phi \frac{d\phi}{(1 + n \sin^2 \phi)\sqrt{1 - k^2 \sin^2 \phi}}.$$

Let us note that when $k \equiv 1$, $E(\phi, 1) = \sin \phi$, and therefore one can consider $E(\phi, k)$ to be a generalization of the function $\sin \phi$.

The Jacobi cosine function $cn(u, k)$ can be defined in terms of $sn(u, k)$ by means of the identity

$$sn^2(u, k) + cn^2(u, k) = 1. \qquad (A.6)$$

The third Jacobi elliptic function $dn(u, k)$ is defined by the equation

$$dn^2(u, k) + k^2 sn^2(u, k) = 1. \qquad (A.7)$$

The integral definition of $sn(u, k)$ makes it clear that $sn(u, 0) = \sin(u)$. Of course, $cn(u, 0) = \cos(u)$ as well.

Besides sn, cn and dn, there are another nine functions that are widely used, and their definitions are given below:

$$ns = \frac{1}{sn}, \qquad nc = \frac{1}{cn}, \qquad nd = \frac{1}{dn}$$

$$sc = \frac{sn}{cn}, \qquad cd = \frac{cn}{dn}, \qquad ds = \frac{dn}{sn}$$

$$cs = \frac{cn}{sn}, \qquad dc = \frac{dn}{cn}, \qquad sd = \frac{sn}{dn}.$$

The derivatives of the elliptic functions can be found directly from their definitions (or vice versa, as in Meyer (2001), where the elliptical functions are defined by their derivatives). For instance, the derivative of $sn(u, k)$ may be computed as follows. In (A.3), suppose that $z = z(u)$. Then,

$$\frac{dF}{du} = \frac{dF}{dz}\frac{dz}{du} = \frac{1}{\sqrt{1 - z^2}\sqrt{1 - k^2 z^2}}\frac{dz}{du}.$$

But from (A.2) and (A.3), we know that for $z = sn(u, k)$, we have $F(z, k) = u$. So, replacing z with $sn(u, k)$ and using $du/du = 1$, we obtain

$$1 = \frac{1}{\sqrt{1 - sn(u, k)^2}\sqrt{1 - k^2 sn(u, k)^2}} \frac{d\,sn(u, k)}{du}$$

$$\frac{d\,sn(u, k)}{du} = \sqrt{1 - sn(u, k)^2}\sqrt{1 - k^2 sn(u, k)^2} \qquad (A.8)$$

$$\frac{d\,sn(u, k)}{du} = cn(u, k)\,dn(u, k).$$

After differentiation to (A.6) with respect of u and taking into account (A.8), we obtain

$$\frac{d\,cn(u, k)}{du} = -sn(u, k)\,dn(u, k), \qquad (A.9)$$

Finally, after differentiating (A.7) and using (A.8) once more, we have

$$\frac{d\,dn(u, k)}{du} = -k^2 sn(u, k)\,cn(u, k). \qquad (A.10)$$

Symbolic computational programs such as *Maple*® or Mathematica® have embedded modules for working with elliptic functions, so these functions can be easily drawn. Graphs of the elliptic sin function sn, cos function cn and function dn are shown in Fig. A.1. We can see that $sn(u, k)$ and $cn(u, k)$ are periodic. We can define their period referring to the definitions above (see A.2)

$$K(k) = \int_0^1 \frac{dt}{\sqrt{1 - t^2}\sqrt{1 - k^2 t^2}}.$$

We can see that $sn(K(k), k) = 1$. Obviously, from the graph, we are also convinced that $K(k)$ is $1/4$ of the $sn(u, k)$ period and that the period of $dn(u, k)$ is $2K(k)$. Of course, this can be checked analytically (e.g., see Woods (1934), p. 368), but this argument satisfies our objectives.

Fig. A.1 Graphs of the elliptic sin function $sn(u, k)$, elliptic cos function $cn(u, k)$ and the function $dn(u, k)$ drawn with $k = \frac{1}{\sqrt{2}}$

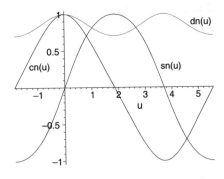

Using the computer program, we can look for a numerical solution, from which we can find $K(k)$, i.e., to solve the equation $\text{sn}(u, k) = 1$. Note that the equation $\text{sn}^2(u, k) + \text{cn}^2(u, k) = 1$ supposes that $\text{cn}(u, k)$ has the same period as $\text{sn}(u, k)$, and therefore $\text{cn}(K(k), k) = 0$.

Now we have an idea of the algebraic and graphic properties of the elliptic functions. In order to "complement" our understanding, let us look at two simple examples—one physical and one geometrical.

Example A.1 (Pendulum) Let the angle of a pendulum swing be denoted by x. Then, it is straightforward to derive that the equation of motion is: $\ddot{x} + (g/l)\sin(x) = 0$, where g is the acceleration due to gravity and l is the length of the pendulum. If we take such units that give $g/l = 1$, the pendulum equation becomes $\ddot{x} + \sin(x) = 0$. Then, we can multiply by $\dot{x}$ to obtain

$$\dot{x}(\ddot{x} + \sin(x)) = 0$$

$$\dot{x}\ddot{x} + 2\sin(x)\frac{\dot{x}}{2} = 0$$

$$\dot{x}\ddot{x} + 4\sin\left(\frac{x}{2}\right)\cos\left(\frac{x}{2}\right)\frac{\dot{x}}{2} = 0,$$

and, by integrating the last equation, end up with

$$\frac{1}{2}\dot{x}^2 + 2\sin^2\left(\frac{x}{2}\right) = c. \tag{A.11}$$

Note also that because of the identity $2\sin^2(x/2) = 1 - \cos(x)$, the last equation expresses the conservation of energy of the motion of a particle with a unit mass. Now let $z = \sin(x/2)$, and therefore $2\dot{z} = \cos(x/2)\dot{x} = \sqrt{1 - z^2}\,\dot{x}$. Then,

$$4\dot{z}^2 = (1 - z^2)\dot{x}^2 = \dot{x}^2 - \sin^2\left(\frac{x}{2}\right)\dot{x}^2 = \dot{x}^2\cos^2\left(\frac{x}{2}\right)$$

$$\dot{z}^2 = \frac{1}{4}\dot{x}^2\cos^2\left(\frac{x}{2}\right).$$

By the first part of the calculation, we have

$$\frac{1}{4}\dot{x}^2 = \frac{1}{2}c - \sin^2\left(\frac{x}{2}\right) = \frac{1}{2}c - z^2 \quad \text{and} \quad \cos^2\left(\frac{x}{2}\right) = 1 - z^2.$$

Hence, $\dot{z}^2 = (A - z^2)(1 - z^2)$, where $A = c/2$. Taking a square root and separating the variables gives us

$$t = \int_0^z \frac{dz}{\sqrt{(A - z^2)(1 - z^2)}} = \int_0^{\sqrt{Au}} \frac{du}{\sqrt{(1 - u^2)(1 - Au^2)}} = F(\sqrt{A}u, \sqrt{A}).$$

That is, we see that the elliptic integrals appear even in the most standard of mechanical situations.

Example A.2 (Ellipse) Let us parameterize the ellipse by the polar angle, which we will denote by t, i.e., $\alpha(t) = (x(t), z(t)) = (a\sin(t), c\cos(t))$, where $0 \le t \le 2\pi$, and $a \ge c$. Then, the arclength integral is

$$L = \int_0^{2\pi} \sqrt{\dot{x}^2 + \dot{z}^2}\, dt = 4\int_0^{\pi/2} \sqrt{a^2\cos^2(t) + c^2\sin^2(t)}\, dt$$

$$= 4a \int_0^{2\pi} \sqrt{1 - \varepsilon^2\sin^2(t)}\, dt,$$

in which $\varepsilon = \sqrt{a^2 - c^2}/a$ is the eccentricity of the ellipse. If we substitute $\sin(t) = u$, then $dt = du/\sqrt{1 - u^2}$, and in this way, we obtain

$$L = 4a \int_0^1 \frac{\sqrt{1 - \varepsilon^2 u^2}}{\sqrt{1 - u^2}}\, du = 4a\, E(\varepsilon).$$

So, we have been convinced once more that the elliptic integrals present themselves even in the most natural geometric problems.

All Jacobian elliptic functions have integrals, given by the formulas below, that can be verified by direct differentiations

$$\int \operatorname{sn} u\, du = \frac{1}{k}\ln(\operatorname{dn} u - k\operatorname{cn} u), \qquad \int \operatorname{cn} u\, du = \frac{1}{k}\arcsin(k\operatorname{sn} u)$$

$$\int \operatorname{dn} u\, du = \arcsin(\operatorname{sn} u), \qquad \int \operatorname{ns} u\, du = -\ln(\operatorname{ds} u + \operatorname{cs} u)$$

$$\int \operatorname{ncn} u\, du = \frac{1}{\tilde{k}}\ln(\operatorname{dcn} u + \tilde{k}\operatorname{scn} u), \qquad \int \operatorname{nd} u\, du = \frac{1}{\tilde{k}}\arcsin(\operatorname{cd} u)$$

$$\int \operatorname{scn} u\, du = \frac{1}{\tilde{k}}\ln(\operatorname{dcn} u + \tilde{k}\operatorname{ncn} u), \qquad \int \operatorname{cd} u\, du = \frac{1}{k}\ln(\operatorname{ncn} u + \operatorname{scn} u)$$

$$\int \operatorname{ds} u\, du = \ln(\operatorname{ns} u - \operatorname{cs} u), \qquad \int \operatorname{cs} u\, du = -\ln(\operatorname{ns} u + \operatorname{ds} u)$$

$$\int \operatorname{dcn} u\, du = \ln(\operatorname{ncn} u + \operatorname{scn} u), \qquad \int \operatorname{sd} u\, du = -\frac{1}{k\tilde{k}}\arcsin(k\operatorname{cd} u).$$

A.2 Weierstrassian Elliptic Functions

Let us consider the elliptic integral of the first kind in the Weierstrassian approach

$$u = \int_z^\infty \frac{dz}{\sqrt{4z^3 - g_2 z - g_3}}, \tag{A.12}$$

in which g_2 and g_3 are arbitrary complex numbers.

It can be proven that the above integral defines z as a unique function of u, denoted as $\wp(u, g_2, g_3)$, in which u is the argument and g_2 and g_3 are the so-called invariants of the function $\wp$.

One should note that if the discriminant $\Delta = g_2^3 - 27g_3^2$ of the cubic polynomial under the square root (A.12) is positive, then it takes real values for real values of z.

Besides $\wp(u)$, Weierstrass introduced another two functions-$\zeta(u)$ and $\sigma(u)$. They are defined by the equalities

$$\zeta(u) = \frac{1}{u} - \int_0^u (\wp(\tilde{u}) - \frac{1}{\tilde{u}^2})d\tilde{u} \tag{A.13}$$

and

$$\sigma(u) = u \exp(\int_0^u (\zeta(\tilde{u}) - \frac{1}{\tilde{u}})d\tilde{u}). \tag{A.14}$$

Of the many interesting properties of the Weierstrassian functions, we will mention only those which have some direct relations to the applications used in the present text, namely,

$$\zeta'(u) = -\wp(u), \qquad \frac{\sigma'(u)}{\sigma(u)} = \zeta(u) \tag{A.15}$$

$$\wp(-u) = \wp(u), \qquad \zeta(-u) = -\zeta(u), \qquad \sigma(-u) = -\sigma(u).$$

References

F. Bowman, *Introduction to Elliptic Functions with Applications* (English Universities Press, London, 1953)

A. Greenhill, *The Applications of Elliptic Functions* (Dover, New York, 1959)

H. Hancock, *Elliptic Integrals* (Dover, New York, 1958)

D. Lawden, *Elliptic Functions and Applications* (Springer, New York, 1989)

K. Meyer, Jacobi elliptic functions from a dynamical systems point of view. Am. Math. Month. Am. **108**, 729–737 (2001)

J. Stillwell, *Mathematics and its History* (Springer, New York, 1989)

F. Woods, *Advanced Calculus* (Ginn and Co., Boston, 1934)

Index

© Springer International Publishing AG 2017
I.M. Mladenov and M. Hadzhilazova, *The Many Faces of Elastica*,
Forum for Interdisciplinary Mathematics 3, DOI 10.1007/978-3-319-61244-7

Printed in the United States
By Bookmasters